D0045006

People and the Sky

Anthony Aveni

People and the Sky

Our Ancestors and the Cosmos

with 65 illustrations, 9 in color

To Davíd Carrasco, colleague and friend,
who has introduced me to so many
interesting people from diverse walks of life.

Half-title *An astronomer of the Renaissance uses a quadrant to measure a star's altitude, Venice 1569.*

Frontispiece *Originating in Egypt and Babylonia, our Western constellations were elaborated upon by the Greeks, then passed on to the limits of Alexander's empire and Islam. They were also transmitted via the Silk route to the Far East. The constellations on this Indian globe, dated 1840, retain a surprising number of recognizable ancient traits.*

First published in 2008 in hardcover in the United States of America by Thames & Hudson Inc., 500 Fifth Avenue, New York, New York 10110

thamesandhudsonusa.com

Library of Congress Catalog Card Number
2007905836

ISBN 978-0-500-05152-8

Printed and bound in Slovenia by MKT Print d.d.

Contents

INTRODUCTION

The Sky, a Legacy Lost

It is almost irresistible for humans to believe that we have some special relation to the universe, that human life is not just a more-or-less farcical outcome of a chain of accidents reaching back to the first three minutes, but that we were somehow built in from the beginning.[1]

A modern cosmologist

At 5:25 p.m. on 21 December, 1880, just as darkness descended, Thomas Edison threw a switch connecting a long string of incandescent lamps on lower Broadway between 14th and 26th Street, Manhattan, to his nearby DC generator. Thus he set the "Great White Way" aglow. The *New York Post* remarked: "Broadway was bathed with a clear, sharp bluish light resembling intense moonlight, with the same deep shadows moonlight casts."[2] The electric light bulb turned night into day. Since that day people in the developed world have rarely found themselves in the dark.

The encroachment of day upon night was already well under way in the gaslight era, when the promise of a better life through improved technology kindled a great mass movement of people from farm to city. By 1900 one out of three U.S. citizens lived in the city. Contrast that number with one in twenty in 1800. (Today it is four out of five.)[3]

Introduced in the 1930s, Daylight Saving Time rolled back nightfall by offsetting the cycle of human activity with nature's diurnal rhythm. Today efficient central heating and air conditioning advertise our preference for the indoor life. Most of us work indoors, shop and dine indoors, run and play tennis indoors. Our eyes spend most of their wakeful hours scanning TV, computer, and movie screens, the contemporary fount of stories that entertain and enthrall us. Often we tan without ever exposing ourselves to the sun. Mindless of the solar course, we look at our watches to tell time and we set appointments electronically by desktop or pocket calendar. If you want evidence of the profound effect technology has had on the way we live, just think what it was like last time you experienced a power outage.

Because we live in a world of progress we tend to dwell on the gains technology has afforded us—instant news, fresh food, the ability to move rapidly from place to place. (Prior to the invention of the internal combustion engine little more than a century ago, the world's fastest traveler rode horseback at 40 km (25 miles) per hour.) But side effects dot the roadway to the easy life. The loss of contact with the natural world is one of them.

Step outdoors for a minute and look around: half your field of view is contained in the hemispherical dome that lies above eye level and is bounded by the 360° horizon that envelops you. If one of technology's downsides is our diminished view of what goes on in nature, the almost total loss of contact between people and the sky is a major blind spot. Today the look of the sky is little more than an imminent weather indicator to most of us. It is rare in our modern world to experience a clear view of a pitch-black, star-studded sky, unobstructed by the glare of urban lighting.

For most of human history the sky *was* relevant. It was alive and present in people's lives. There was rhythm in the sky—there was pattern and hierarchy. The pristine perfection of the firmament offered a compelling medium for human connections. Throughout the seasons people found meaning in the dance of the cosmic denizens that populated it.

Technology isn't the only reason we lost the link between the sky above and everyday life here below. A change in the basic way we think about the natural world contributed to our detachment as well. Our social estrangement from heaven began in the Renaissance and continued through the late 18th-century Enlightenment. A new ideology portrayed nature as a world apart from human affairs. The sky, like the earth, came to be regarded as inanimate matter; it possessed neither soul nor human meaning. We mathematized and objectivized the universe—made it an entity to be described and understood "as it is" and "for itself," purely in quantitative terms. We sought to explain its workings strictly according to abstract natural laws that we believed lay embedded within its structure, hidden laws that only await human discovery. Our new faith held that these laws, properly applied, were capable of leading to predictions that could be made manifest and tested, not through prayer or some other form of dialogue with imagined transcendent powers, but rather via repeated dispassionate observation and precise measurement. We invented science.

In the late 16th century Copernicus placed the sun at the center of the solar system and opened the human imagination to a hitherto unrealized sort of universe—a vast realm populated by countless stars for us to probe. Half a century later technology placed the telescope between eye and sky. In the capable hands of Galileo this novel instrument revealed that the planets—bright lights we had long followed as they coursed among the stars in the belief that they had been put there for the sole purpose of influencing our fates—were really worlds of commensurate rank with our own. Once seen only as tiny points of light on the fixed dome of the sky, the stars were suddenly transformed into blazing suns at incomprehensible distances. In a flash of history an idea and an instrument shrank our terrestrial habitat to a microcosm.

Today the fruits of scientific technology and enlightened ideology relegate us to a cold, indifferent, expanding universe which started with a bang some fourteen billion years ago and shows no sign of limiting its extent either in space or in time—a universe we take for granted, attempt to cope with, but still only rarely take the trouble to gaze out at.

So what exactly are the losses that come with the profound gains in knowledge we've acquired of the world around us? What aspects of the everyday intimate contact between how we live and what happens in the sky have disappeared from the sphere of human concern? How and why did our great transformation happen? These are the questions we shall explore in *People and the Sky.*

A Harris Interactive Poll of U.S. adults conducted in 2006 listed science and firefighting as the most prestigious professions. (There has been a drop of 17 percent for scientists since 1977.) Lawyers (17 percent) and journalists (14 percent) headed those of lowest prestige. Had such a poll been conducted in 6th-century BC Babylon, 7th-century AD Yucatan, or among the contemporary Batammaliba of West Africa or the Desana of the South American Amazon, I am confident that "the one who watches the sky" would have headed the list.

I have chosen "walks of life" as a medium for telling the story of the multitude of diverse ways stargazing has contributed to the development of human culture. Through this medium it is easy to see how we have become estranged from these imaginative ways of knowing the natural world.

What better way to begin to turn our eyes to the night sky than by telling stories? Storytelling was once a major activity associated with skywatching. To recount all the stories would take up many thick volumes, but there are some common denominators among tales from different cultures. In Chapter 1, "The Storyteller's Sky," I have selected a few choice tales from around the world that depict how our forebears once understood the heavens. The stories I have chosen focus on a few basic questions: Why do we believe we came from the sky? Exactly how did we get here from there? Where did we get the idea of putting gods in heaven? What is so special about "up"?

People the world over have always managed to find star patterns there. We call them constellations, and we can trace their origins back to the very beginnings of civilization. In Chapter 2, "Patterns in the Sky," I look at why we invent constellations, focusing in particular on the ones that seem to be almost universally recognized in a variety of cultures—like Orion, the Pleiades, and the Big Dipper.

Who *needs* the sky? Today no administrator, clerk, accountant, lawyer, doctor, laborer, teacher, or technician would argue that being able to identify stars and constellations, knowing the phase of the moon, or the hour and position of sunrise has any relevance to their job description. But had you been able to ask the ancient hunter, farmer, sailor, or astrologer—all quite respectable livelihoods in civilizations past—you would have heard a very different response. In *People and the Sky* I devote a chapter to each of these noble, cosmically grounded professions.

"The Sailor's Sky" (Chapter 3) tells about the master navigators of Polynesia, who charted courses across a seemingly limitless watery world without any knowledge of latitude and longitude; they took their bearings from the stars, and developed, with little technology, sophisticated navigational systems to explore far-flung new lands. The polar Inuit accomplished much the same, with vast fields of snow and ice replacing the ocean. Scholars once thought nomadic and semi-nomadic hunter-gatherers would have had little interest in charting sky movements; but as we learn in my fourth chapter, "The Hunter's Sky," we have evidence today that the !Kung and Mursi of Africa, among others, relied on signals in the sky for their very survival and sustenance.

"The Farmer's Sky" (Chapter 5) was a timepiece. The seasonal arrival and departure of particular constellations told when to plant and when to harvest. From the ancient Greek poetry of Hesiod to the contemporary testimony of Indonesian rice planters, we learn the agrarian secrets of success acquired by carefully charting time cycles in the heavens.

"As it is in heaven" is a phrase that aptly reflects how the sky mirrors life. Social cohesion and kinship ties follow the way of the sky in cultures as diverse as the Inca of Peru and the Pawnee of the U.S. Midwest, whom we visit in Chapter 6, "The House, the Family, and the Sky." The Gilbert Islanders of Micronesia, the Warao of Colombia, the Batammaliba of Togo, and the Navajo of the U.S. Southwest are among the people whose homes we must enter in order to see for ourselves how they conform to patterns derived from the sky.

In my view too little attention has been paid to the role of the sky in placemaking on a grand scale. In "The City and the Sky," the seventh chapter, I take up this issue. Just as the sky once defined the location and arrangement of the family household, so it also functioned in setting up the shared space of the wider social community. I have chosen four diverse examples to illustrate that divination—discourse with the gods—was a major motive in urban design. The Skidi Pawnee arranged their village in accordance with a template based on the constellations that passed overhead; the stars also ordained the order of seasonal rites and the social conduct of the Pawnee peoples. On a much larger scale, the plans of 2,000-year-old Teotihuacan and its habitué, Tenochtitlan (today's Mexico City), offer urban spaces adapted to the practice of human sacrifice to the gods of war and fertility—the victims paying the debts of the people to the deities and thus enabling the sun to remain on its course and the life of the community to continue. Likewise, Etruscan cities and temples of 6th-century BC Italy were precisely situated to facilitate prayer to the gods of the sixteen directions of the horizon, whom the high priests invoked through the entrails of a sacrificed sheep. Finally, ancient Beijing, a city with a written history, offers detailed clues regarding the lives of the Chinese diviners, who planned the urban environment in accordance with the situation of the perpetually visible polar constellations; the focus of the city structure was the Purple Palace with its royal inhabitants, most prominent among them the emperor who resided at the pivotal point corresponding to the Pole Star.

If the chapters on the house, the village, and the city resonate a personal note, this is because they were all developed out of the interdisciplinary field of archaeoastronomy, an area of study I helped establish. As we shall see, this science offers an ideal example of the fresh insights to be gained by looking down the cracks between the disciplinary floorboards of the house of knowledge. Archaeoastronomy, the study of ancient astronomy through evidence from both the written and the unwritten record, challenges the material medium we usually rely on to discover truth about the past—the record that appears on the printed page—and exposes its fallibility and incompleteness. The importance of archaeoastronomy lies in the recognition that we can understand a good deal more about the ancient science of the sky by examining evidence that transcends the written word—evidence from art, sculpture, stories of creation, and beliefs and social customs. My eighth chapter, entitled "The Ruler's Sky," follows on from the survey of the city, showing how the urban scene legitimized dynastic rule by publicly proclaiming the power of the elite to be derived from heavenly ancestors.

No book on past perceptions of the night sky would be complete without a chapter on "The Astrologer's Sky" (Chapter 9). From Egypt, Mesopotamia, and Greece to Mesoamerica and Peru, this highly respectable profession expanded the universal maxim "as above, so below" into a highly complex mathematical system capable of yielding precise predictions of the positions of the sun, moon, and planets—all for the purpose of foretelling their influence on humanity. We shall discover that, without the groundwork laid by the astrologers of old, there surely would be no scientific astronomy today.

The last two chapters deal with how the view of the heavens developed in the West. They are, of necessity, slightly more technical than the rest. "The Timekeeper's Sky" explores several examples of human attempts to capture time in the written record, and ends with the great technological explosion that engendered precise scientific timekeeping. Herein lies the great irony: the skillful creation of complex mechanisms to capture time ended up harnessing those who sought to control it. But as we shall see, other cultures, such as the ancient Maya of Yucatan, though they did not invent telescopes, astrolabes, and chronometers or leave written evidence of their work, were nonetheless quite capable of developing similar complex systems.

Lastly, "The Western Sky" tells the story of how and why we in the modern West came to understand the world around us so differently from our forebears. I deal especially with how the Classical philosophers created a parting of ways from the common roadway of human society, towards a peculiar way of thinking of the sky as a world that functions quite apart from human interaction.

People and the Sky, then, is more than a culture-by-culture survey of astronomy designed to herald the diverse accomplishments of each. That story has been told many times over. Nor is it a call to revert to our earlier relationship with the sky—there is indeed no turning back. Rather, *People* offers a journey into a lost world, a world where our forebears were more intimately tied to, and dependent upon, nature. In our attempts to view the sky through diverse human eyes and to comprehend the enormous differences between these old ways of knowing nature and our own, we shall discover that we still share with our ancestors a common curiosity and quest for celestial knowledge.

CHAPTER 1

The Storyteller's Sky

This is plain—that there in Teotihuacan, they say, is the place; the time was when there still was darkness. There all the gods assembled and consulted among themselves who would bear upon his back the burden of rule, who would be the sun.[1]

A 16th-century Spanish chronicler of the Aztecs

The universe was created some fourteen billion years ago. In the flash of an instant an essence inexplicably sprang into existence. This colossal, violent explosion happened everywhere, causing a very rapid expansion that we still witness today. The expansion cooled things down and the essence separated into what we now recognize as energy and matter. Eventually matter came to dominate the universe. Its microscopic structure formed immense gas clouds and galaxies of stars, some surrounded by planets. In at least one fortunate instance that we know of, people gaze out toward the jettisoned remnants hoping to reconstruct the story of creation. But there may yet come another creation once the universe ceases to expand, falls in on itself, and implodes anew … and so on, and so on. (Today the odds are against such a scenario.)

This brief (and admittedly oversimplified) account of modern science's version of creation hardly does justice to the familiar Big Bang hypothesis; nevertheless, it is enough to set up an interesting contrast with other creation tales—stories told by diverse people from around the world. Let's recount a few of them.

In the beginning there were two: Tonacatecuhtli and Tonacacihuatl, Lord and Lady Sustenance; they lived in the thirteenth heaven, the highest of all the heavens. It was they who gave birth to the gods: to Red and Black Tezcatlipoca, gods of the smoking mirror who could foresee in the reflections of the stars (Plate 8) any change or conflict that might happen in the future; to Quetzalcoatl, the feathered serpent god of all that is creative, who gives balance to our lives; to Huitzilopochtli, Lord Hummingbird on the left, god of sun and war; and to Tlaloc, god of rain.

People say that Black Tezcatlipoca made the first creation—the first "sun." He called it the "sun of earth." His handiwork was a world populated by a race of clumsy giants who unintentionally destroyed everything around them. Fortunately Quetzalcoatl intervened, striking down Black Tezcatlipoca and placing him in the sky in the form of the jaguar constellation—the one we call the Big Dipper. Then he sent forth jaguars to devour the giants. They say that today if we dig into the earth and look carefully we can find the bones of the giants of the "sun of earth."

Then Quetzalcoatl made his own universe, the "second sun." He called it the "sun of wind." But Black Tezcatlipoca returned from the sky and defeated Quetzalcoatl, who, along with his sun, was carried off by the winds. Monkeys were the original inhabitants of the second sun. They still swing from the trees as a reminder of that ancient age.

Next Tlaloc tried his hand at it. He made a third creation, the "sun of rain." But Quetzalcoatl destroyed that world with a rain of fire—the same sort of rain we still see today when the great volcano erupts and spews ash over our city. The water goddess Chilchiuhtlicue, the lady of the jade skirt, then tried her hand at creation. She fashioned the fourth creation, the "sun of water," only to be swept away by a huge flood that changed the world's inhabitants into fish and destroyed the mountains that hold up the sky.

After four failed attempts the gods became aware that their dissension and rivalry over how to create the best possible world was ultimately responsible for its destruction. And so, for the next creation, the gods joined forces. Black Tezcatlipoca and Quetzalcoatl transformed themselves into giant trees and set out to raise up the sky again. But first they were required to slay a horrible earth monster named Tlaltecuhtli, an indescribably fierce creature who looked like a cross between a snake and a caiman. To do battle the courageous brothers transformed themselves into serpents. They grappled with the monster, finally managing to grip it at opposite ends. They tore it apart. Out of the lower half of the body they fashioned a new earth and from the upper portion they made a new sky. For their heroic deeds the brothers were made Lords of all the Heavens by the rest of the gods. You can see the road they use to cross the star-studded sky over which they rule—it's called the Milky Way. Finally, to take care of their new world the gods fashioned people, and they created animals and maize for their sustenance. Having set the stage, all the gods gathered at Teotihuacan (see my epigraph) for their final creation—the fifth sun, the "sun of time."

It is morning twilight; the eastern horizon lights up. Day is about to dawn, but there is no sun. "Who will carry the burden?" the gods ask themselves. "Who will take upon himself to be the sun, to bring the dawn?"[2] The arrogant Tecuciztecatl, Lord of Snails, steps forward: "I shall be the one," he says. But as he moves toward the horizon to cast himself into the flames he is singed by the intense heat and he quickly draws back. He tries a second approach, then a third and a fourth, but each time he fails to gather enough courage to take the plunge. He backs off, as all the other gods remain silent. Then Nanauatzin, the leprous Pimply One, humbly attired and wearing a simple paper headdress, steps up; he closes his eyes and slowly walks forward. Without hesitation he hurls himself into the fire. His body crackles and sizzles. The gods look around in anticipation of the rising sun. "Where will it come, they ask?" They look south, then north, then

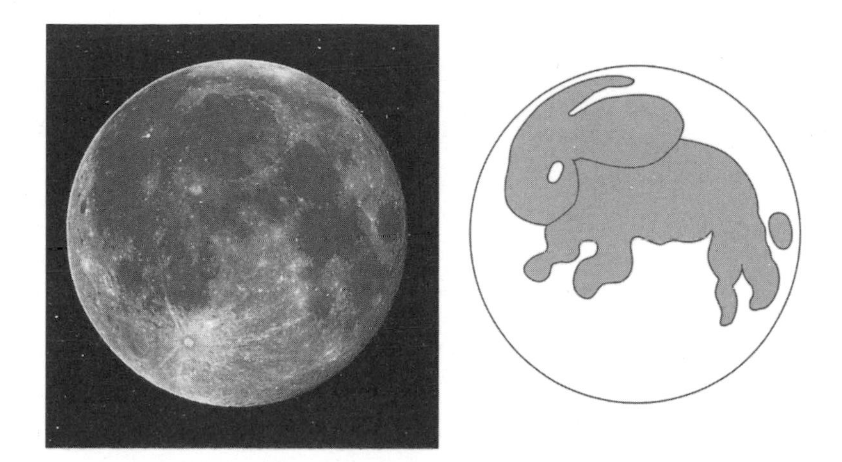

1 *The Rabbit in the Moon: If we look at the lunar disk we can see the same shape that skywatchers from both North America and the Far East have identified as a rabbit (the sketch on the right helps). Did the idea of a lunar rabbit spring from one culture and spread throughout the world, or did it occur independently to those who looked to the heavens for an understanding of the world around them? Perhaps we will never know. The Aztecs say their gods threw a rabbit at the face of the moon to dim its luster so that it would no longer rival the sun.*

west, and finally east. There Tonatiuh, the newly created sun of time, rises. At first he is flaming red; he moves only in stutters and jerks. "How shall we live?" say the gods to themselves, realizing that to start up time, to make it move smoothly, they all must sacrifice themselves. And so they do.

Emboldened by Nanauatzin's deed, Tecuciztecatl casts himself into the flames and emerges just as brilliantly in the sky, in the guise of the moon. To darken his face one of the gods comes running. He hurls a rabbit at the moon's face to dim it so that it will not rival Tonatiuh. You can still see the rabbit there today (Fig. 1, Plate 7).

> "Will they perchance both together follow the same path?" wonder the gods. But the moon holds fixed until the sun finishes spending the day doing his work—which is why they say that after the sun disappeared in the west, "the moon undertook the night's task, working all night."[3]

Now Tonatiuh needs nourishment if he is to continue his daily course across the sky from east to west. And so it becomes the sacred duty of the people, the custodians of the earth, to follow the example set by their creators. Only they can ensure that the great cosmic drama that unfolded at sacred Teotihuacan will continue, by offering their own "precious water"—human blood; only they can avert the perpetual darkness that will befall the fifth creation, the sun of time, the sun of movement, should his light become extinguished. So goes the Aztec story of creation.

We call all such stories 'myths' (the Greek word *muthos* means story or fable) and I believe myths are too often misunderstood. Our Big Bang seems to avoid that label. Writes astronomer Carl Sagan: "The chief difference between them and our modern scientific myth of the Big Bang is that science is self-questioning, and that we can perform experiments and observations to test our ideas."[4] So, can our story of creation be true and theirs not? If we look beyond the falsehood and fantasy we associate with myths, we arrive at a deeper measure of truth for those who tell them. Myths don't just entertain or frighten us. They are important for their informative life lessons and moral values.

The Aztec myth is all about gods who behave like people. Conflict over their attempts to create a more perfect world arises out of their lack of cooperation. They succeed only when they discover strength in unity. The swinging monkeys, erupting volcanoes (Popocatepetl still smokes on the periphery of modern Mexico City—Tenochtitlan, where the ancient Aztecs lived), bones of giants (they turn out to be mammoth remains), and celestial bodies of the story—told and retold—reminded people that they must make huge sacrifices to keep the world going. Why? Because all higher powers need someone to worship them, to provide sustenance for them so that they can continue to enable life on earth to exist. People repay their gods by continually re-enacting the ultimate sacrifice they believe once happened at Teotihuacan, the place where time began. The blood of human sacrifice was part of a reciprocal relationship between the people and the gods that kept time in motion. So, the Aztec sky myth of creation was more than just a story. It was the core of an ideology, a system of religious beliefs that held a proud people together for nearly two centuries before the Spanish invasion of the New World. Myth has meaning. There is truth in it.

Like the Aztecs, the Maya of highland Guatemala say that it took a great drama, a great performance, to bring about the emergence of "all the sky-earth":

> by the Maker, Modeler,
> mother-father of life, of humankind,
> giver of breath, giver of heart,
> bearer, upbringer in the light that lasts
> of those born in the light, begotten of the light;
> worrier, knower of everything, whatever there is:
> sky-earth, lake-sea.[5]

In the beginning there was not a single human, animal, fish, or tree, no canyon, meadow, or forest. There was only the sky above, and the sea below. Then the maker-modeler, the mother-father talked about how to create the "sowing and the dawning." First the creator tried it by word—speaking the words: "earth—let it be this way," and the mountains rose from the water, separated, and formed the streams and valleys. Thus were created animals, birds, and finally, after many attempts that met with failure, people, fashioned to carry out the demanding task of taking care of their new environment. But the gods became anxious, for the light of dawn had already begun to appear in the east, yet there was no sun.

There were imposters to be sure—like the boastful Seven Macaw, who showed off his glittering gold teeth as he fancied himself the great luminary, and his crafty son Zipacna who had killed the Four

2 *Hero twins in a Maya story of creation subdue a boastful bird who pretends to be the sun. One twin lures him to his lofty perch (note the beckoning arm extending from the base of the tree) while the other shoots him with a blowgun. As a reminder of the deed the bird is placed in the sky where he becomes the Big Dipper. Far to the south (at the bottom of the world tree) lies the scorpion constellation.*

Hundred Boys, the gods of fermented beverages, and cast them into the sky where they became the Pleiades. These false gods—along with the evil gods of pestilence who ruled the Underworld and were bound to ruin any legitimate creation—needed to be dealt with.

And so from the sky descended the hero twins, Hunahpu and Xbalanque. They dispatched Seven Macaw by knocking him out of his perch in the cosmic tree with a blowgun (Fig. 2). As a perpetual reminder of the evils of pride and self-centeredness, they cast him into the sky where he became the constellation we call the Big Dipper. After doing away with the forces of evil on the surface of the world, the twins descended into Xibalba, the Underworld, to play a game of ball with the Lords of Xibalba—Blood Gatherer, Bloody Teeth, Jaundice Master, Pus Master, and Bone Scepter. Employing superior skills of cunning and trickery, the twins defeated those bearers of maladies and disease, whose only concern was to make life unbearable for any race that would inhabited the earth.

Finally, when all had been put in order the two boys ascended:

> straight on into the sky, and the sun belongs to one and the moon to the other. When it became light within the sky, on the face of the earth,

> they were there in the sky. And this was also the ascent of the Four Hundred Boys [the Pleiades] killed by Zipacna. And these came to accompany the two of them. They became the sky's own stars.[6]

Last to be created were the people—descendants of the gods. They were placed upon the earth to care for it, to keep the sun dawning forever by their actions, by their rituals of remembrance of an ancient past.

Like the Aztec story of creation, the Maya myth begins in the sky. Its principal characters are also human-like gods who do not necessarily agree on how to create the world, but who finally succeed in making a less perfect world better. The myth lauds the heroic qualities and it draws on the observation of nature to provide clues to, and reminders of, a storyline that tells people who they are and why they exist. True, the Aztecs and the Maya are both a part of ancient Mesoamerican culture; consequently they may have shared ideas about stories of creation. But there are tales from the other side of the globe with strikingly similar attributes.

Marduk is the hero deity of the Babylonian myth of creation. He was born of male Apsû, the Begetter of Gods, and Tiamat, "she who gave birth to them all,"[7] long before heaven above and earth below had been named. Now, the children of Apsû and Tiamat were far from perfect. Like terrestrial children, they were disorderly offspring who ran around the house of heaven making noise and behaving mischievously. When they disturbed his sleep Apsû threatened to do away with his children for being so clamorous; however, he was outsmarted and killed by members of the second generation. His spouse, transformed into a fire-breathing monster, sought revenge for her husband's death. She marshaled an army of monsters, vipers, terrifying dragons, mad dogs, and scorpion men—indescribably vicious, sharp-toothed *lahamu*—whose bodies are filled with poison instead of blood. Her son Marduk (Zeus to the Greeks, Jupiter to the later Romans, Thor to the still later Nordic people) realized that the only way to establish order would be to confront Tiamat with an army of his own. In a climactic scene, as violent as any Hollywood high-tech-effects film, the two approached one another and engaged in hand-to-hand combat (Fig. 3). The narrative continues:

> The lord spread out his net and enmeshed her;
> The evil wind, following after, he let loose in her face;
> When Tiamat opened her mouth to devour him,
> He drove in the evil wind, in order that [she should] not
> [be able] to close her lips.
> The raging winds filled her belly;
> Her belly became distended, and she opened wider her mouth.
> He shot off an arrow, and it tore her interior;
> It cut through her inward parts, it split [her] heart.[8]

3 *Heroic deities who slay cosmic monsters are a common theme in many creation stories. Here Marduk, the patron god of Babylon, subdues the water goddess Tiamat, who has morphed into a winged monster. The sky myth is a reflection of the Mesopotamian taming of the waters via the mastery of irrigation.*

Out of his mother's vanquished body Marduk created the world. He split her in two and cast half her body into the sky; the other half became rooted in the surface of the earth.

Next Marduk set about carefully fashioning the contents of heaven. He made stations for the great gods—the signs of the zodiac. He divided up the year, and for each of the twelve months he set up three constellations (see the discussion of the zodiac in Chapter 2). He opened gates of heaven on both sides, and made strong locks to the left and to the right. Finally he turned his attention to creating the moon, the ornament of the night:

> Monthly without ceasing go forth with a tiara.
> At the beginning of the month, namely, of the rising o[ver] the land,
> Thou shalt shine with horns to make known six days;
> On the seventh day with [half] a tiara.
> At the full moon thou shalt stand in opposition [to the sun], in the middle of each [month].
> When the sun has [overtaken] thee on the foundation of heaven,
> Decrease [the tiara of full] light and form [it] backward.[9]

That narrative is written on a clay cuneiform tablet that once stood in the public square of 1st-millennium BC Babylon, the great metropolis between the Tigris and Euphrates in what today is Iraq. The cosmological creation myth is called *Enûma Elish* (meaning "when above," after the first words in the cuneiform text) and it was passed on orally for a thousand years before it appeared in the written record. The myth celebrates Marduk, leader of all the gods, who became the patron deity of their great city, and it describes how he brought together the forces of nature (the various deities named in the epic) to create the world. But *Enûma* tells us even more than the Aztec and Maya stories of creation about the people who championed great Marduk and what was on their minds.

Like those of us who live near earthquake faultlines and flood-prone coastal and riverine zones, anyone who resided, worked, and prayed in ancient Babylon would have appreciated just how menacing nature can be. Such a person would have understood why every move the gods of nature made merited the closest attention. What is really behind *Enûma* is a battle for control of the universe between the forces of the sky and earth deities. In ancient times, the myth tells us, the gods lived intermingled in harmony, just as the sweet waters of the Tigris and Euphrates, represented by male Apsû, blended with the salt water of the Persian Gulf, personified by female Tiamat. At this tenuous juncture between two watery worlds the earthly abode was built up out of the gradual silting that still takes place in the delta (today southern Iraq). The war that broke out among the gods typifies what life can be like in a land where nature's balance of power is never really secure.

In reality, the agrarian world in the land between the two rivers experienced a seasonal cycle punctuated by tension and disorder. In winter the steady rain in the mountains in the north produces episodes of flooding in the valleys to the south. In spring the wind howls southward along the narrow fertile strip between the rivers to begin the drying-out process. Then suddenly the summer sun turns on the world and mercilessly attacks it, his blazing light unrelentingly baking the landscape into drought. So it is no surprise that inhabitants of the volatile fertile crescent would perceive the creation of the present world as the outcome of a battle between the forces of earth and sky.

The *Enûma Elish* is an ancient theogony (a story about the origin of the gods). It tells a tale that resonates with child and grandparent alike—a wild and violent adventure story of how the present world came about as a result of generational battles between parents and children. Out of the maelstrom emerges the invincible Marduk. He slays the earth monster and from her body, slit in half and blown outward by the force of the wind, he creates the world essentially as we know it, with land situated below a starry domed sky and floating upon a flat ocean. And so the gods of the next generation seek order. To maintain that order Marduk creates in the sky the pristine environmental timers, the sun and the moon, and Jupiter (Marduk's star) in the zodiac to retrieve astrological omens about the future. Later in the myth he creates people for the express purpose of serving the gods, that they might remain contented in their present condition.

The people are charged to tame the chaotic waters by developing a system of irrigation to sustain their agriculture, and to control the floods that would otherwise inundate them. They must also build a great *ziggurat*, or pyramid, dedicated to Marduk. And they must write the tablets that tell his never-to-be-forgotten story, displaying them in the center of their city to reaffirm and disseminate the truth of it.

Performing the *Enûma* in Babylon became part of a great New Year's festival. In periodic rites in honour of the gods, Babylon's good citizens re-enacted certain aspects of the myth over and over again, lest it be forgotten. Marduk is the king of the gods and his earthly manifestation is the role-model of the king. *Enûma Elish* teaches us that, in a violent world, you succeed only by meeting force with counterforce. The only way to achieve order is by force, and this necessarily entails violent action—the only viable means by which Marduk could bring about the present world. Such an ideology made a lot of sense in the urban imperial state that was Babylon in the middle of the 1st millennium BC.

Tension, conflict, and resolution underpin most lasting creation myths. And resolution usually comes with a pricetag—the service the people must offer their gods by conducting rites that re-enact the creation. But the heavenly element in the story—splitting open the monster who threatens to destroy the world and fashioning the sky out of its remains—resonates worldwide.

What is it about us that we need to relate the stories of our lives to

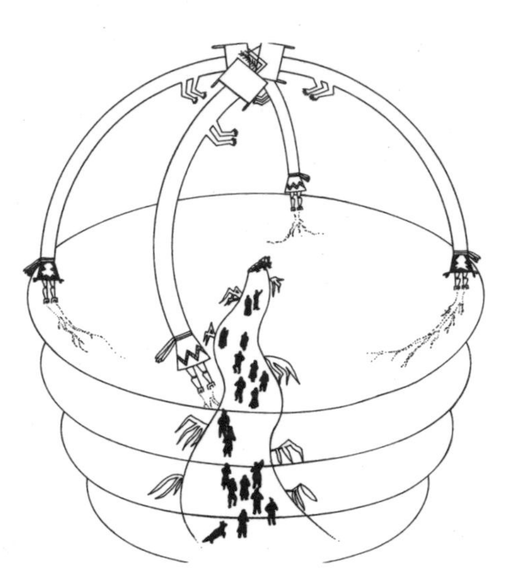

4 *Creation for the Navajo consists of a series of emergences by people through holes in layers of a hierarchically arranged world. With each successive emergence people become more domesticated until finally they are civilized enough to reach the earth's surface, where they are destined to become the custodians.*

the sun and the stars? To keep our thoughts in order, to keep things from happening to us, a Navajo girl once answered an inquiring anthropologist. *Diné Bahané* ("a telling about the people"), the colorful myth of Navajo creation, also dwells on how to attain harmony and balance with the forces of nature. It is filled with the same sort of generation and gender conflict that we find in the other sky stories I have summarized. The core of the Navajo narrative is an adventure story about cycles of emergence from one world into the next in a vertical hierarchy (Fig. 4).

Our story opens with a precise description of a four-part universe with oceans on each side, into which four streams flow outward from the center, in the cardinal directions. The dwelling places of the races who inhabit each quadrant are designated by their colors: white in the east like the dawn, blue in the south like the day, yellow in the west like the twilight, black in the north like the night.

The people who lived in the four quarters fought among themselves and with one another. The insects—called the Air Spirit people of the south—were the worst offenders. They committed adultery. Their actions so offended the other worldly inhabitants that they were told to find some other place to live. Said the others: "We do not want you here." And so the Air Spirit people were banished from the world. They traveled many days upward across the hard dome of the sky until they saw a hole in the heavens near the eastern horizon. Passing through it, they emerged into a second world, one inhabited by the blue-headed Swallow people who lived in blue cone-shaped houses scattered across a broad blue plain. The two races became friendly and at least for a while they lived in harmony, until one of the insect males became a little too friendly with the Swallow chief's wife. The chief confronted the Air Spirit people and lectured them: "We treated you as kin, and this is how you return our kindness? Leave our world immediately."

Then the Air Spirit people traveled upward to the top of the second world's sky. They found an opening at the southern horizon. It led to a third world, a yellow-colored world inhabited by yellow Grasshopper people. But once again the intruders wronged the chief by violating social customs. "Get out of here," intoned the chief, and once again the Air Spirit insect people were forced to take flight. Flying to the top of this third world they encountered yet another sky, where they chanced to enter another world through a hole at its western horizon. Now they were confronted by an even more bizarre fourth world. Its colors alternated at regular intervals between white, blue, yellow, and black. Unlike the other worlds, it seemed that no one lived there. There was neither sun nor moon, only a snow-covered peak in each direction. After extensively exploring this exotic new land the insect people encountered a strange race of people who lived in upright houses in the ground. They cut their hair and groomed themselves. They married. They cultivated the land and gave their visitors corn and squash to eat. They got to know the insect people and they taught them to cleanse themselves, to mind their ways, to become civilized.

Once the insect people bathed and the women dried themselves with yellow corn meal and the men with white corn meal, they were magically transformed into First Man and First Woman. But even real people commit incest and real married people also quarrel. So, even in the fourth world disorder overcame harmony; however, this time the transformed people had enough sense to seek a new path on their own. After a lengthy series of adventures they managed to find their way through the last aperture—the one that leads to the fifth world, the one we know today. There they finally learned to live in harmony with nature and with one another.

To light their fifth world—to make it brighter than all the rest—First Man and First Woman made the sun to light up the night:

> First they fashioned the sun. They made an object round and flat, something like a dish, out of a clear stone called *tséghádi'nídínii* or rock crystal, as *Bilagáana* would call it today.
>
> They set turquoise around the edge of this dish. And just beyond the turquoise they placed rays of red rain. Beyond that they placed bars of lightning. And beyond those they placed shimmering swirls."[10]

Then they made the moon, another object round and flat like the sun. They made it out of *tsésq'*, or rock-star mica. To adjust and refine the cosmos to suit the people's needs, like Marduk they also created the months with their cycle of moon phases.

But where should the sun rise and how should it set? The east wind beckoned to them to bring the new light to his edge of the world. There they placed the sun in the care of a youth who had spread sacred soil in the fourth world below. The moon was carried by an old man who brought the soil to the youth. They say that all who die will be placed in their care as fair exchange for the work they do here in the fifth world. But the nighttime sky of the fifth world was still too dark. More lights were needed up there for those who wished to travel by night, especially when the moon did not shine. And so First Man and First Woman fashioned the stars. These they also molded out of rock-star mica so people could see them shining in the sky with their extra light.

Then *Áltsé hastiin* the First Man sketched a design on the ground, so that he could work out a plan for lighting up the heavens. Once he was satisfied with his scheme, he began to carry it out.

Working very slowly and very carefully, he placed one fragment of mica in the north. There he wished to have a star that would never move. By it those who journeyed at night could set their course.

Then he placed seven more pieces of rock-star mica. Those became the seven stars we now see in the north.

Next he placed a bright piece of mica in the south. Likewise, he placed one in the sky to the east. And he put another one in the sky to the west. He did so very carefully and very thoughtfully.

So it was that he slowly built several constellations. For he wanted the results of this work to be perfect.[11]

Fragments of four stories told by four different peoples. Each gives us insight into what really mattered to those who created them. Whether the information in these myths can be validated by scientific testing—the way we can measure redshifts in the spectra of distant galaxies to validate the Big Bang—doesn't really matter. Thousands of similar tales from people past and present all over the world tell of universes created by gods like themselves, who behave like artisan craftspersons, or of universes formed by the hatching of a cosmic egg, or made by the dismemberment of a giant primeval monster. Action heroes ascend and descend, emerge and re-emerge through hierarchically arranged layers of existence to which only they can bring order and meaning by gradually transforming the unfamiliar into the familiar. The harmony we seek both in nature and society—a harmony typified by the creation of the sun, moon, and stars whose pristine motion constitutes the clockwork of the universe—can only be achieved and maintained through repetitive human action.

As cosmologist and Nobel laureate Steven Weinberg laments in the epigraph to my introductory chapter, it is the absence of human involvement that makes modern science's sky story so different from the rest. In the tales I have recounted, people participate in their cosmologies: they play an active role; they engage in a dialogue with supernaturals; they sacrifice; they conduct rituals that retell the story of creation, a story that reminds them all of the parts they play every day in a great human drama set upon a cosmic stage. Our story, with its central theme of a Big Bang that happened billions of years ago and brought everything into existence, includes the microcosmic seeds that would eventually become us. But the story we tell is about a colossal cataclysmic drama over which we really have no control. We are simply not part of the cast of characters. We have written ourselves out of the script. Our creation narrative comes equipped with no clues that pertain to the search for human meaning. We are mere bystanders in a history that began in an instant at a point in time inconceivably long ago.

True, our story is different and the reasons why it is different are encased by a history we'll explore in a later chapter. For now we might better be served not by singling ourselves out but rather by looking at ways in which all stories of creation resemble one another. There is a common denominator that unites us all—the desire for order arrived at through a quest for pattern. In the next chapter we'll explore the universal habit of the search for patterns in the sky.

CHAPTER 2

Patterns in the Sky

A priest of ancient Sumer years ago, while worshipping his God of heaven ... conceived of certain figures which to him seemed proper to express forever the love and adoration felt by man for Him, Who made and governs sky and earth and sea, and all the waters underneath the earth.[1]

A 19th-century historian

Recently I visited the caves of Camuy in central Puerto Rico. Our guide took great pleasure in pointing out fanciful resemblances to plants and animals of the shapes of stalactites and other cave features: "Doesn't that look like the face of a man in profile?" "There's a mountain lion." "I call this 'broccoli'." And "That looks like a strip of bacon, doesn't it?" Tourists eagerly aimed their cameras in the direction of each likeness, to document the perspicacious guide's disclosures. When I returned from my trip a paper appeared on my desk for me to evaluate. It was about "earthworks" on the surface of Mars in the shape of a bird and a human head (shown in profile)—alleged signs of life on an otherwise barren planet.

Sharing interpretations of observed shapes on cave walls and satellite photos reminded me of the summer days of my adolescence, when my friends and I would lie on the beach a few blocks from the neighborhood where we grew up. Especially on days when puffy cumulus clouds raced across a deep blue sky, we would spend hours conjuring up cloud pictures, watching them morph from animal to human profile and back again, sometimes even forming a late model sedan, a bicycle, or other familiar images from our everyday world. Often we would criticize one another for not being able to see the obvious.

Was it the same for our ancestors when they watched the stars file across the firmament at day's end after the cloud picture show evaporated? Did ancient shepherds of the Middle East, with little more to do at night than tend their flocks, muse about the resemblance of the Big Dipper to a wagon, or Orion to the frontal view of a man? I have never been able to subscribe to the popular notion that Orion got its name because anyone who looks at it automatically sees a hunter. In fact I doubt that anyone who gazes at the early winter southern sky will even recognize, except perhaps for the three belt stars that line up, the entire star pattern we call Orion as separate from the adjacent constellations of Eridanus, Taurus, Canis Major and Minor, and Auriga. (To this day when I look at Auriga I fail to see a man riding on a stagecoach carrying a bunch of baby goats.)

Indeed what we call Orion, or portions thereof, was once known as the Fire Sticks to the Aztecs, the Tortoise to the Maya, and the One-Legged Man to the Carib Indians. To test any doubts I had, I once handed out star maps to my astronomy students and asked them to find star patterns. But first I removed all the coordinates and traces of connector lines between stars. The sheets were entirely unlabeled—just a field of randomly spaced dots. Fewer than 5 percent of the class identified any part of a recognized constellation, though a few sharp-eyed individuals did succeed in recognizing the bowl and handle of the Big Dipper.

The historical record strongly indicates that naming patterns in the sky was not a matter of discovery, but rather the result (as the epigraph reminds us) of a deliberate framing of figures that possessed religious or mythic significance, perhaps an attempt to commemorate the glory of the gods who created the world. Sharp-eyed eagle, powerful bull, brave lion, fiery dragon, pure virgin, nurturing bearer

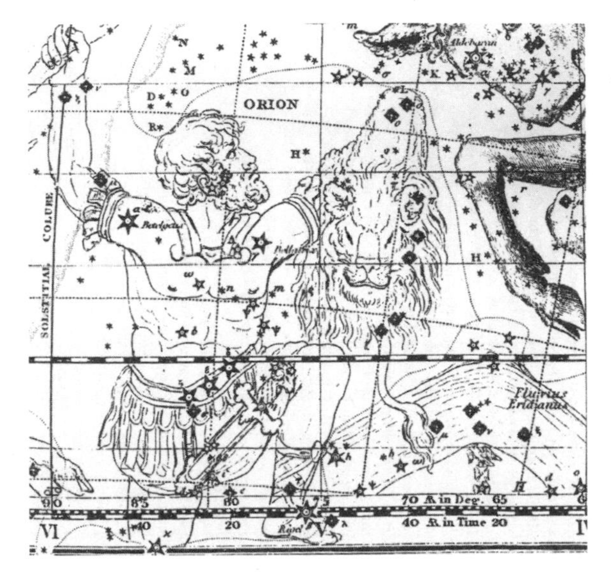

5 Do we all see the same star patterns in the sky? Orion the hunter in Western lore is easily identifiable by his belt and sword. Bright red Betelgeuse marks a shoulder, lustrous blue Rigel a foot. He defends himself with cloak and clubs from charging Taurus the Bull.

of water, deadly scorpion, king's crown and chariot—all these are figures that might spring from the imagination of an individual situated close to royalty, one who might have wished to project such imagery onto the space where worshippers cast their pleading eyes for whispers of what the future would bring.

To judge by their names, our constellations descend from 3rd-millennium BC Sumeria. They make their first recorded appearance on boundary stones and cuneiform texts in the 7th century BC, and in the contemporary epic tales of Homer and Hesiod. The route of descent then passed through Classical Greece, Islam, and Renaissance Europe.

Our constellation names are Arabic. For example, Orion the hunter (Fig. 5), originally was termed Al Jabbar, the giant. Bright red Betelgeuse, Ibt al Jauzah, is the "armpit of the central one" (in other designations it is called the "shoulder," "arm," or "right hand of the giant"). Rijl Jauzah al Yusra, shortened to Rigel, is the bright blue star that marks his left leg. The closely gathered line of three bluish stars of Orion's belt were together regarded as the golden nuggets that lay at the center of the constellation. Each had its own designation. Mintaka, on the right, means "belt," while Alnilam in the middle is the string of pearls set at the center of the belt; last to rise, Alnitak is the girdle. Up in the other shoulder lies Bellatrix, the only prominent star in Orion that does not commonly bear an Arab designation, though on old maps it is labeled Al Murzim or Mirzam, the "roaring conqueror." Less luminous Saiph indicates Orion's fainter right leg. Actually Saiph means "sword" and was originally intended to mark the tip of the weapon that hangs from his belt. The brightest star in the faint and fuzzy handle of the sword (which houses Orion's great nebula) is Na'ir al Saif, a literal translation of "brightest one of the sword." Al Maisan (Meissa), apparently the result of an erroneous juxtaposition with a star in neighboring Gemini, was once the head of Al Jauzah, or Ras al Jauzah. And finally the tidy little string of stars along the upheld arm above the right shoulder that represents the sleeve of the garment he wears were collectively called Al Kumm ("sleeve").

Despite the experience of my students, many different cultures have recognized the same patterns in the sky. In China the names of constellations appear inscribed on oracle bones of the Shang dynasty as early as the 14th century BC. The *Rig Veda*, a Hindu hymn dating from the 2nd millennium BC, refers to the constellations, as does the writing in royal tombs of 16th-century BC Upper Egypt. In the Americas, the Maya, Incas, and Aztecs fashioned star patterns after things that mattered to them. To illustrate the diversity of sky-oriented cultures and the details of the stories that go with their star patterns, let's keep our gaze focused on our winter constellation of Orion.[2]

The Greek version of the story of Orion identifies him as a demigod, a son of Poseidon, god of the sea, born of a daughter of King Minos of Crete. While visiting the court of an Aegean island king, Orion, emboldened by an excessive intake of fine wine, tried to ravish the royal princess. As punishment the king blinded him, took away his power, and sent him wandering. A benevolent god took pity on Orion and gave him a servant to guide him to the place where the sun rose, so that he might once again walk across the sea—a power once vested in him by his father. When he arrived, the sun god restored his sight. Ultimately Orion found refuge on Crete, where he became a celebrated game hunter. But self-indulgent excess, this time encouraged by Artemis, goddess of the wild, and perhaps an adrenaline rush from the excitement of the kill, once again got Orion into trouble. When he boasted that he would slay every animal on the face of the earth, Gaia, the earth goddess, became alarmed. She delivered a scorpion, who did away with the emboldened hunter by stinging his heel (another version of the tale has Artemis launching the scorpion following Orion's amorous advances upon her).

The story of Orion is a cautious ancient Greek reminder that anyone guilty of *hubris*—in this case boasting of divine prowess—earns retribution from the gods. This is why both hunter and scorpion were memorialized by being placed, safely opposite each other, in the nighttime sky. More than this, key points in the story can be cued by various aspects of the constellation; when Orion vanishes following sunset in late spring he is blinded; his vision is restored when he returns to the night sky in the middle of summer.[3]

An ancient Chinese constellation myth that dates from the 12th century BC casts Orion and Scorpio in the roles of quarreling brothers destined to prowl opposite sides of the sky, one in the winter, the other in the summer. Historians suspect that Zhou rulers used them to teach moral lessons, parallels for the two previous dynasties who failed because they constantly antagonized one another.

At least one of the New World readings of the constellation also interprets it as a male figure known as Epietembo, the One-Legged Man (Fig. 6). Contemporary Carib people of northern South America tell the adventure-love story of a newly married young woman who is tempted by a lover, a young man who has taken on the form of a tapir just to be near her. He promises to assume human form once they reach the eastern horizon if she will agree to ascend into the sky with him. As she collects wood for the fire, she secretly

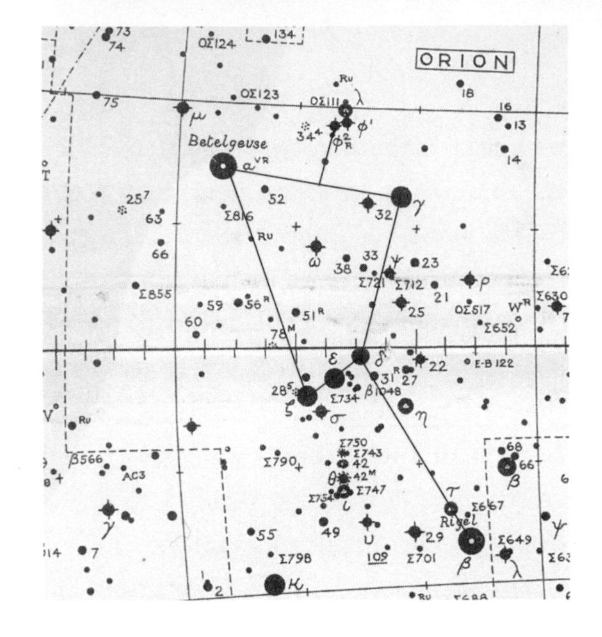

6 *Among the Carib of the New World Orion is a One-Legged Man, thanks to his dissatisfied spouse, represented by the Pleiades, who has hacked off the missing limb.*

deliberates the proposal, while her husband, Orion, engages in gathering ripe avocados nearby. When Orion descends from one of the trees, the young woman grabs her ax (which has been enchanted by her tapir lover), cuts off his leg, and takes flight with her new beau. But Orion recovers and, supported by his crutch, he seeks out the whereabouts of his lost love, casting a trail of stones from the avocados that sustain him on his journey.

After a long and tortuous quest, Orion finds the couple making love, cuts off the head of the tapir, and implores his wife to return. Instead she elects to pursue the tapir-lover's spirit into the heavens. Undaunted, the husband follows in hot pursuit. You can still see them there: the woman (the Pleiades), next to the tapir's head (the Hyades)—his bright red eye is the star Aldebaran—and of course, the jilted husband, Orion, his remaining lower limb marked by bright Rigel, following close behind.

The eternal love triangle, a marriage broken by seduction: there are plenty of morally based themes in this tale to discuss around the hearth, and plenty of visual props in the celestial winter triad that consists of Orion, the Pleiades, and the Hyades. But, as in the Old World myth of Orion, there is a seasonal as well as a symbolic framework in telling stories through star patterns. Act I: the encounter with the tapir-lover happens in the dry season when avocados ripen and wood is collected; Act II: the husband's frantic search occurs during the rainy season, when trails run cold; Act III: during the planting season he finds the amorous duo, after the seeds of the avocado fall on the ground. The constellations act out the myth of fertility on the sky stage. Thus the Pleiades rise before dawn in mid-June, followed by the Hyades, and then by Orion.[4]

When the Lakota of the upper midwest in the U.S.A. gaze at the region of Orion, they see not the whole figure of a man but rather a hand. Orion's belt makes up the wrist, and his sword outlines the thumb. Rigel is the tip of the index finger, and the star β Eridani, borrowed from the constellation adjacent to the west, serves as the end of the little finger (Fig. 7). The configuration of the Lakota Hand constellation may be different from our familiar Western star pattern, but the lesson about the chief who lost it is the same—how *not* to behave.[5]

A young chief proposes marriage to the daughter of the head of a neighboring tribe. She agrees but, as in the Carib story, there are conditions. The chief must recover the hand of her father, which was torn away by the spirit Thunder people because the old chief was selfish and refused to make divine offerings. Taking up the challenge, the young chief travels from place to place among the Black Hills of South Dakota, acquiring special powers from friendly spirits that enable him to escape from the Thunder people

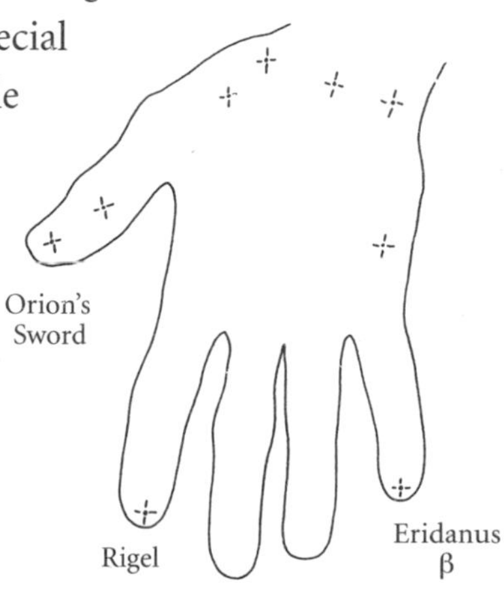

7 The belt and sword along with Rigel of our Western constellation of Orion are all that are recognizable in the Lakota constellation of "The Hand" of an Indian chief.

should they capture him. After a lengthy ordeal he succeeds in recovering the lost hand and he returns it to the chief. This in turn wins him the chief's daughter as a wife.

The struggle with the Thunder people is necessary to restore order, and—transgressions forgiven—there is a happy ending; but viewer-listeners also understand the seasonal timing behind the placement of the Hand in the sky. The Hand disappears at the end of nature's fertility cycle. It reappears in the autumn—a sign that the sacrifices offered during the summer solstice ceremony, which re-enacts the myth of the Hand, have been effective. Renewed life will be possible in the next seasonal cycle. The old chief represents the previous cycle, and his successor (by marriage in this instance) is the new cycle—the one that replaces the old through the generative power inherent in the human hand. The daughter is the fertile earth mother, while the son they ultimately bear signifies the life forms that emerge from her in the new year. Indeed, Orion has many meanings. (For other depictions, see Figs. 8 and 9.)

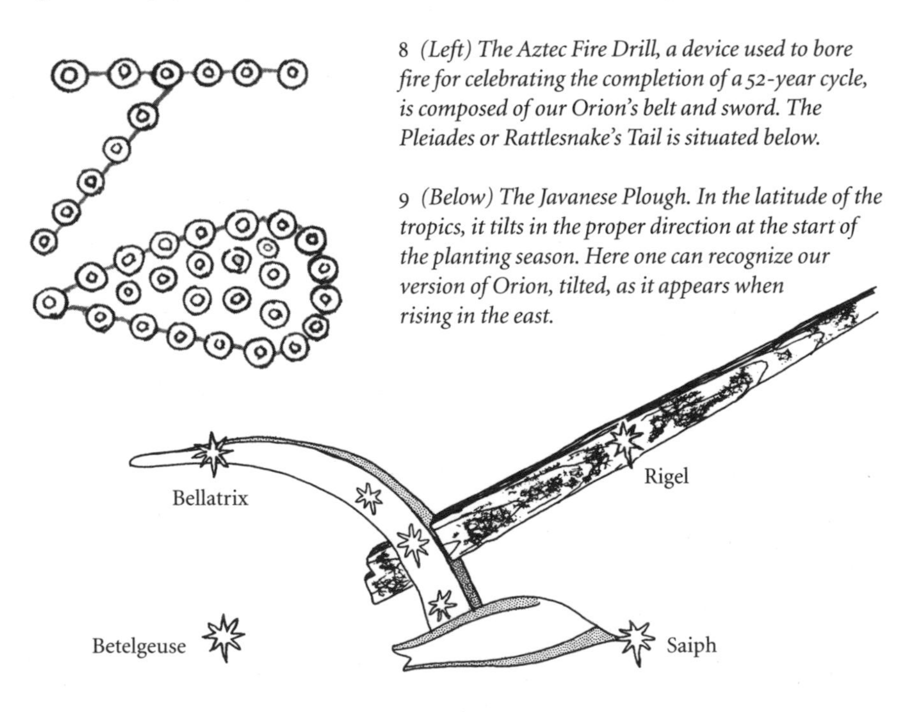

8 *(Left) The Aztec Fire Drill, a device used to bore fire for celebrating the completion of a 52-year cycle, is composed of our Orion's belt and sword. The Pleiades or Rattlesnake's Tail is situated below.*

9 *(Below) The Javanese Plough. In the latitude of the tropics, it tilts in the proper direction at the start of the planting season. Here one can recognize our version of Orion, tilted, as it appears when rising in the east.*

Some patterns in the sky are more special than others. The zodiac (meaning "circle of animals") is a continuous band of stars that passes all the way around the sky. The constellations of the zodiac—there are a dozen such star patterns in our Western version (Fig. 10)—have received a lot of attention because they mark the celestial corridor through which move the sun, moon, and the planets that are visible to the naked eye (Mercury, Venus, Mars, Jupiter, and Saturn). A 6th-century BC cuneiform text from Babylonia reads: "Month 2 Night 23: Mercury is below the star Beta in Gemini by 2½ degrees," and "Month 12 Night 24: Mars is below the star Beta in Capricorn by 2½ degrees."[6] The motive for making such statements seems to have been to reckon the planetary, lunar, and solar cycles—that is, to devise a calendar based on the observation of nature. (As we shall learn in later chapters, some cultures went well beyond practical lengths to calculate these time loops.)

If I know what happened during previous cycles in the world below, based on data acquired from repeated observations of events in the world above, I shall be better able to prepare for what lies ahead. Past experiences coupled with memory offer me a key to the future. This is the essence of what is written in Hesiod's *Works and Days*, an oral poem from the 9th century BC about the farmer's use of the sky (see Chapter 5 for details). But a deeper underlying meaning attaches to the "as above, so below" motto: far beyond functioning as time indicators, sky events in most cultures of the world—the Age of Reason in the West being the notable exception (see page 225)—have been thought to exert an influence on the behavior of all forms of life that dwell upon our planet. Statements such as those quoted above became "if … then" statements, with a clause added about human concern: "If Mars approaches the Scorpion (then) there will be a breach in the palace of the prince."[7] Astrology, the art of predicting such influences, employs precisely this motto. It is a complex subject that deserves a chapter of its own (see Chapter 9).

Western tradition has it that the planetary wanderers traveled "the way of Anu," as the ancient Sumerians called the zodiac. The bright stars that lit the cosmic highway were called "counselor gods" or "consultants." They surveyed what happened in their own zones and advised the gods about future undertakings. The northern region of the zodiac, near the Tropic of Cancer, constituted the way of Enlil or Bel, lord of the earth, whereas the southern part, the region of the Tropic of Capricorn, was the way of Ea, god of the waters. Anu's territory, appropriate to the god of the sky, encompassed the bulk of the zodiac. Dated from approximately 1100 BC, the texts known as the *Mul Apin* list the constellations associated with each zone. Each deity was also assigned a political territory: Elam (to Anu), Akkad (to Enlil), Amurru (to Ea), and so on.

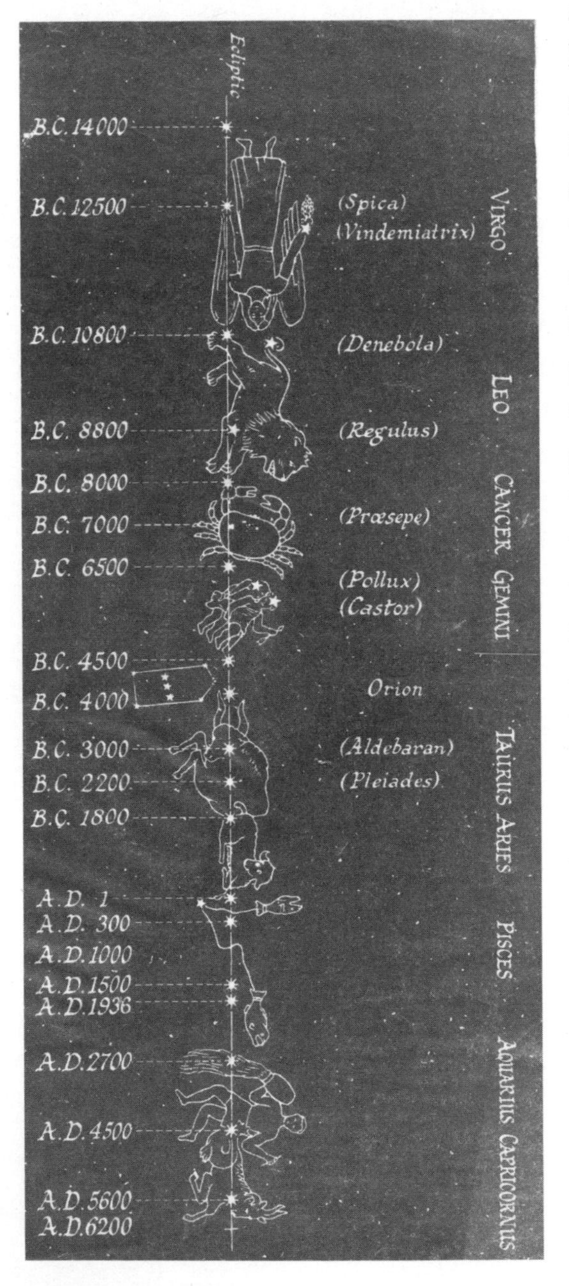

10 *A portion of the Western zodiac as we know it, showing the place of the spring equinox sun through the ages. Our zodiac is descended largely from the Greeks.*

The earliest zodiac was probably lunar. The *Mul Apin* text specifically names the eighteen constellations that stood in the path of the moon, as opposed to the familiar twelve of the solar zodiac (Fig. 10). In ancient Babylonia, Aries the Ram occupied the first 30° of the circle of the zodiac. It designated the position of the sun on the vernal equinox, 20 March. (Over the 3,000 years that have since elapsed that position has backed up into the constellation of Pisces.) Next, the sun eclipses Taurus the Bull , the 30° to 60° zone which corresponds to the dates between 19 April and 19 May; then come Gemini the Twins, 60° to 90°, and so on, the sun passing successively through Cancer the Crab, Leo the Lion, Virgo the Virgin, Libra the Scales (which has the distinction of being the only inanimate star pattern in the Western zodiac). Next there's Scorpio the Scorpion, then Sagittarius the Archer. During the last quarter of the year, which corresponded to the winter rainy season, running from the December solstice to spring equinox, the sun traversed the watery constellations: Capricorn the Sea Goat, Aquarius the Water Bearer, and Pisces the Twin Fish. Because the moon travels farther north and south among the stars than the sun, it covers more territory. (Additional constellations in the original list of eighteen include the Pleiades, Orion, Perseus, and Auriga. Pisces was once subdivided and an extra constellation was recognized between Pisces and Aries.)

Use of the number twelve probably gave rise to the practical idea of dividing the year into months, based upon correlating the phases of the moon to the constellations whose annual disappearance in the west took place when the sun entered each one. For the sake of uniformity the special twelve were made equal and each was assigned a 30° zone of the 360° ecliptic, the circle that bisects the zodiacal band, and which also serves as an approximate count of the number of days in a year.

History documents the trading of ideas as well as goods across the Eurasian continent, so it is no surprise to learn that the Chinese, too, had a zodiac; but scholars who have examined it in detail actually find few parallel identifications (Fig. 11). The Chinese named 28 star patterns distributed in a band centered not on the ecliptic but rather along the celestial equator; these were used to chart the moon. The most familiar animals of the solar zodiac are Rat, Hare, Dragon, Sheep, Monkey, Dog, Boar, and so on. With the exception of the Ox (Bull), these are nothing like our Lion, Twins, Virgin, and Scales. Some investigators think these creatures were chosen to fit the seasons of the year during which they were most active, an idea that makes sense to me given the tendency to relate constellations to seasonal activities such as fertility. Thus, the bellicose Rooster, which marks the full moon that falls in our month of October, signaled the time after the harvest when preparations were made for conducting raids and warfare. And the Monkey (November) gives birth at that time of year, in tune with the full moon of that month.

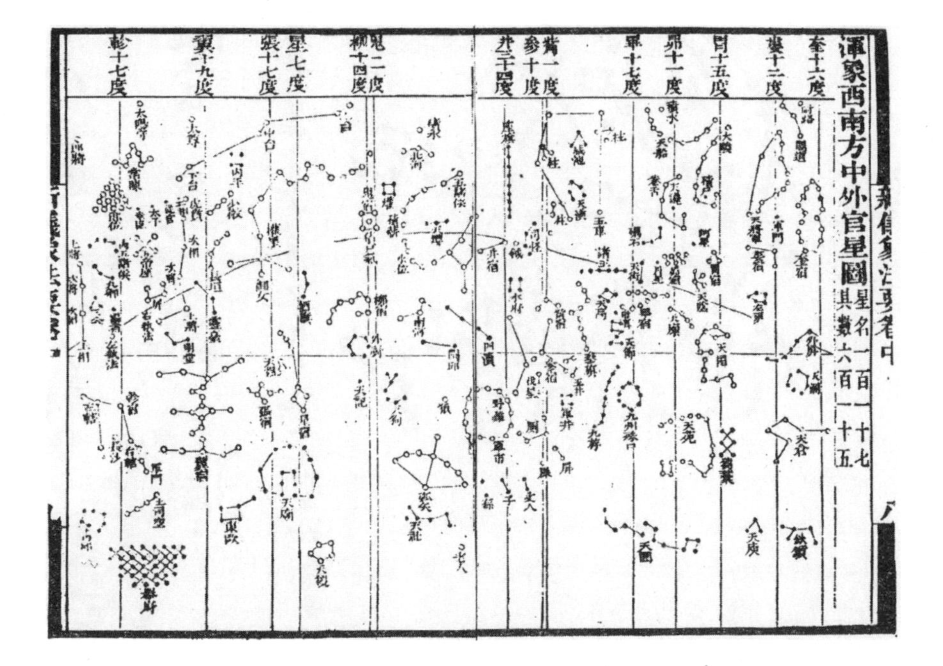

11 *A portion of the Chinese zodiac, 11th century. Fourteen of the 28* hsiu *(lunar mansions) are marked along the top. The ecliptic is the wavy line, the horizontal midline the equator. Only the perspicacious eye can detect any portions of familiar Western constellations.*

A cross-cultural exchange of ideas certainly took place between Egypt and the Classical world, but once again, when we turn to the Egyptian zodiac, I believe independent invention offers a better explanation. It is best documented in "astronomical ceilings" found in tombs and temples throughout the Nile valley. The oldest appears in the Tomb of Senemut (dated to approximately 1550 BC), confidant of Queen Hatshepsut and architect of her magnificent temple in Thebes, capital of Upper Egypt. It consists of a list of decans, stars, or groups of stars whose heliacal rising, or first annual morning appearance, defined the 36 ten-day periods that made up the 360-day year (five days were later tacked on at the end to give a better match between the measurement of time and the actual length of the solar year). Constellations displayed on the ceiling include Osiris in a boat (Orion) and Isis in a boat (the star Sirius, whose heliacal rising heralded the annual flooding of the Nile and the marked New Year).

More famous, and more reminiscent of the sort of zodiac familiar to us, is the ceiling of the Temple of Hathor (goddess of love and heaven) at Dendera, dated to the 1st century BC and shown in Fig. 12. The constellation map is circular in shape, boasting a Crab constellation, a Scorpion lady, a Lion whose tail is held by a goddess (the latter two relate to our Scorpio and Leo), and Isis holding an ear of corn (a possible reference to Virgo). There is also a Hippo deity, and a Bull's Foreleg.[8] The upper register of an oblong diagram on the portico of the temple lays out the constellations of the zodiac in a continuous band.

In the 1st millennium AD, and an ocean away in Mesoamerica (a name given to the region that encompasses culture areas in Mexico and northern Central America), the Maya developed forms of artistic expression, architecture, sculpture, and a calendrical science viewed by many to have been on a par with that of the Classical world (see Chapter 10). They too devised a zodiac (Fig. 13). That the Maya zodiac, a virtual parade of animals starkly similar to those that com-

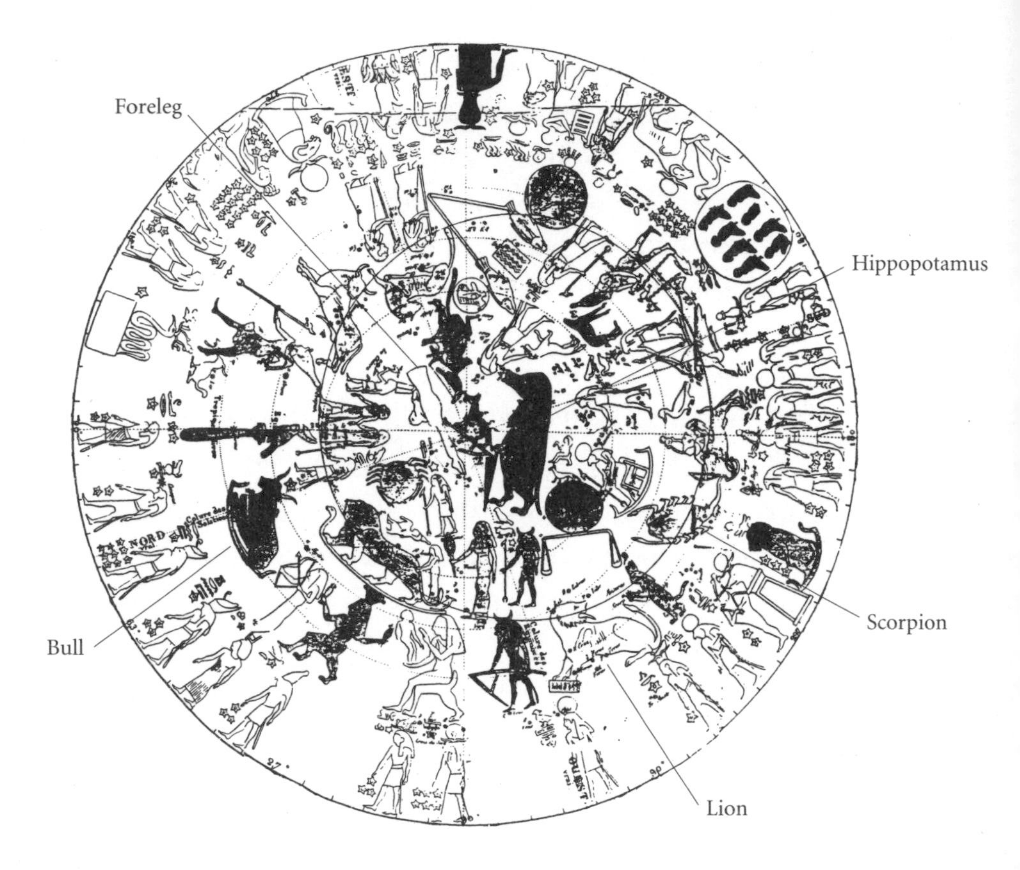

12 *The zodiacal ceiling at the Temple of Dendera, Egypt, 30 BC. Foreleg and Hippopotamus mark the pole. The zodiacal constellations, which are largely Greek (a few are labeled), lie along the inner circle.*

prise our own, might have arisen independently surprises even those who have studied it in detail. The most telling example of it is painted on pages 23 and 24 of the Paris Codex, a written Maya document dated to the 14th or 15th century AD, clearly a time before Spanish contact. This badly eroded book, which unfolds accordion-style, was found abandoned among a pile of soot-covered papers in a chimney corner of the Bibliothèque Nationale in Paris in the 1860s. (We will have occasion to discuss some of the startling astronomical details in other surviving codices in Chapter 10.)

The Paris Codex depicts a sequence of thirteen animals hanging with clamped jaws from Maya sun hieroglyphs situated below a continuous band. Judging from the style of the rest of the codex and from the content of adjacent page 22, the band represents the body of *caan*, the two-headed sky serpent. The animal parade continues across the lower band, which is undecorated. The most easily identifiable creatures in the Maya zodiac are rattlesnake—the rattle is visible at its extremity; tortoise; scorpion; a pair of birds—the first is probably a vulture; and serpent. Less certain are frog; deer; death head; and peccary. Are these open-jawed celestial animals devouring the planets which enter their respective zodiacal houses? Eclipses and planetary conjunctions are often portrayed in this manner in other Maya works of art.

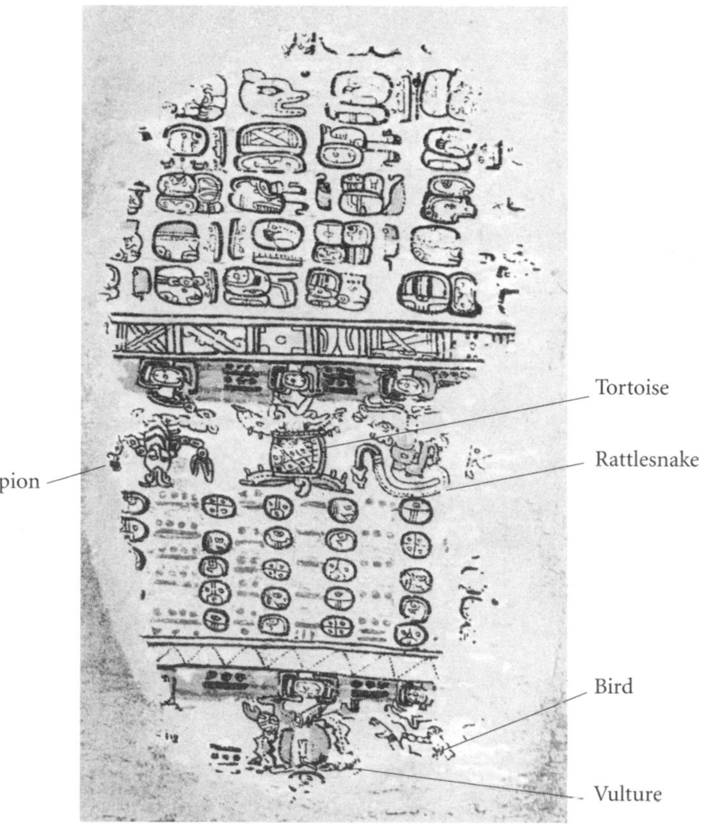

13 *Part of the Maya zodiac in the 15th-century Paris Codex. Here the parade of animals dangle from decorated bands representing a sky serpent.*

The calendar that occupies the remaining space on the zodiac pages implicates the moon. Each day name entry is separated from the next by 28 days, the approximate number of days when the moon is visible. Since 13 x 28 = 364, the approximate number of days in a year, the sun also appears to be involved. Like our modern corrupted lunar year (twelve months of approximately 30 days) such a calendar would soon slip out of phase with the actual motion of the moon, which would require a correction. This could be where the numerals eight and double-eight, written in green in dot-and-bar notation on the sky band between the pendent figures, come in. To date, no one has explained the numerals to any general satisfaction. Other numbers of Maya calendrical significance may be implicated.

The Maya zodiac has also been rendered in a mural painting showing a scene of surrender in the aftermath of a battle, and on at least two sculpted friezes on Maya ceremonial buildings. Scholars have had problems aligning the Maya animals with the parade of seasonal star patterns that appear in our own zodiac. An ingenious hypothesis suggests that the difficulty may lie in our insistence on understanding the zodiac as a *continuous* parade of animals. What if the constellations that comprise the Maya zodiac are not laid out in linear fashion, strung across the sky along the ecliptic as our culture views them? Rather, if we draw upon the Maya emphasis on horizon-based astronomy, it may be that star groups were schematically arranged at the skyline during morning and/or evening twilight. Specifically, the constellations may be arranged in the Paris table in alternating pairs so that, for example, when the first one is positioned just above the east horizon, the second lies just above the west, some 160° away. This model has succeeded in arriving at matching definitions for practically all of the constellations in all the Maya texts.[9]

The Barasana who dwell far up the Orinoco River in Venezuela are among the cultures in the contemporary New World who possess a zodiac (Fig. 14). This time the evidence comes from an inquiring

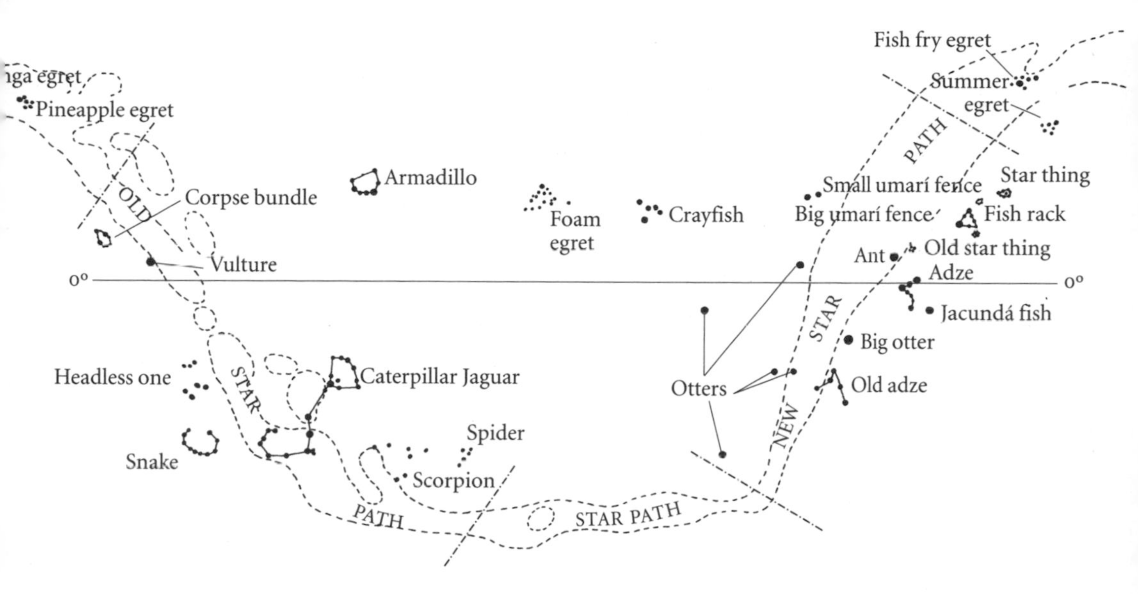

14 *The Barasana, a tribe of South American Indians, conceive of an imaginative two-part zodiac made up largely of riverine creatures that dwell in their habitat.*

anthropologist,[10] who learned that the Barasana divide their zodiac into two halves. There is an Old Path, centered on our Scorpio, consisting of all things associated with death and decay. Among its denizens are a poisonous snake and spider, the headless corpse of an eagle, a vulture, and—oddly enough—the body of a woman who has been stung to death by wasps. In stark contrast, the New Path contains only things of sustenance, such as a fruit arbor, a fish smoking rack, Jacundá fish, big and little otters, and crayfish. The lead constellations of each path control the corresponding portion of the season, the Pleiades or Star Woman marking the beginning of the New Path, while the Caterpillar Jaguar, a jaguar with a snake for a tail (the equivalent of our Scorpio) connotes the Old Path.

Star Woman is in charge of the seasons and agriculture. She appears at dusk on the eastern horizon to mark the end of the rain and the beginning of the swidden (slash-and-burn) agriculture, and she disappears in the west after sunset in April to end the dry season and bring on the heavy rain that fertilizes the manioc. Caterpillar Jaguar does just the opposite. He dominates the skies of the April to November rainy season (rising and setting heliacally at those times). Scarcity of food and an increase in disease are attached to his presence; but fortunately, they say, he also brings the Caterpillar people, a species of moth that pupates during the rainy season, when it falls down from the trees above to become a staple in the local diet during these otherwise hard times. In ceremonies today men still dress up in bright feather patterns and painted body stripes to imitate the caterpillars. They dance to celebrate the movement of the seasons, for which they believe their actions are responsible.

In some places the constellations of the Milky Way serve as the zodiac. This is particularly true in the southern hemisphere, where the band of stars, glowing gas, and dark interstellar dust that make up the plane of our Galaxy are particularly prominent. Contemporary Andean people find their principal constellation patterns in the dark clouds (*yana phuyu*) of the southern Milky Way, where star clouds and the dark spaces between them are densest (Fig. 15). The dark streak between the constellations of Scorpio and the Southern Cross represents a llama suckling her calves. The brightest stars, α and β Centauri, mark her bright eyes. Other animals in the stellar parade across the firmament include fox, partridge, toad, and a great anaconda. Like other zodiacs, the Milky Way served as a functioning environmental calendar.[11]

Thus the intervals during which many of these sky creatures make their appearances correspond to periods when their terrestrial counterparts are most active. For example the partridge, called *yutu* (the constellation in the shape of a coalsack adjacent to the Southern Cross), makes its heliacal rise early in September and disappears from view in mid-April. A Spanish chronicler tells us that in ancient Inca times (about 500 years ago) the beginning of this period corresponded with the time farmers needed to guard their crops against these voracious birds. Likewise, real toads re-emerge from the earth just about the time the black cloud toad in the sky first clears the horizon, bringing rain with him. When the 16th-century chronicler Juan Polo de Ondegardo wrote that "all animals and birds on the earth had their likeness in the sky in whose responsibility was their procreation and augmentation,"[12] one wonders whether he realized that he was dealing with a highly organized astronomical system of tracking environmental time. More than an aid to memory, these constellations were thought to hold the power to bring good or bad fortune to the hard-bitten farmer of the high Andes.

For the contemporary Boorong of Victoria, southeast Australia, the chain of dark clouds that make up the southern Milky Way are thought to be the smoke of the fires of the *Nurrumbunguttias*, the old spirits associated with the sun who inhabited the earth in the previous

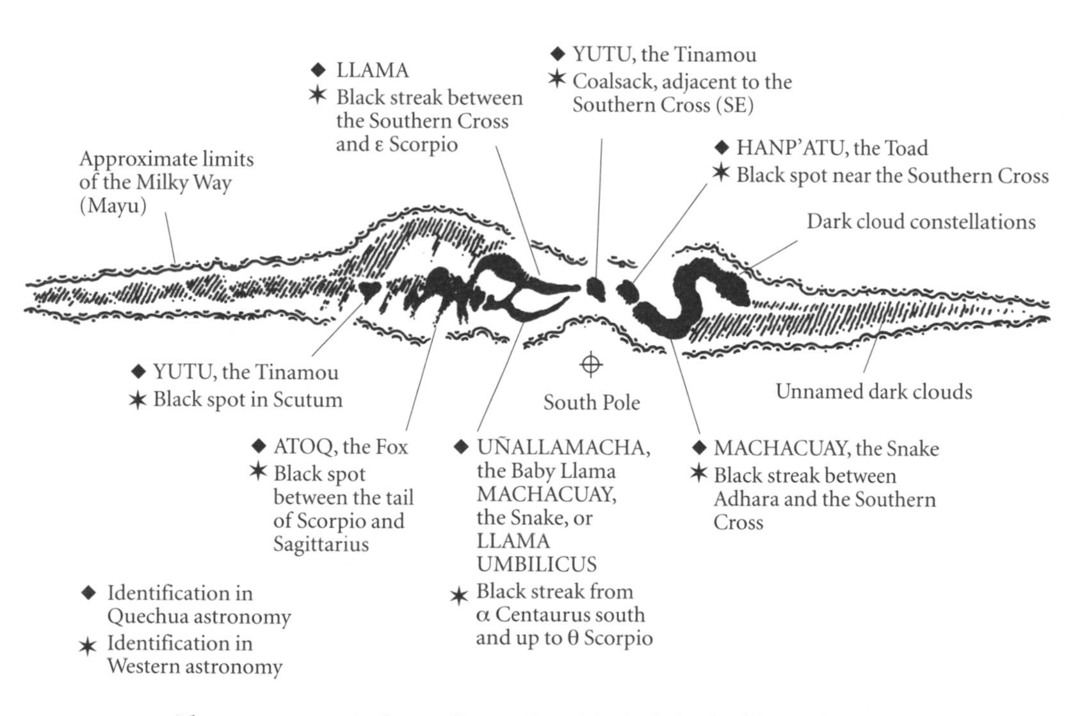

15 *The contemporary Andean zodiac consists of the dark clouds of the Milky Way, which are far more prominent in the southern Milky Way than in the north.*

creation.[13] Bright patches along the southern Milky Way complement the black clouds sandwiched between curtains of closely gathered stars. The Magellanic Clouds—mini-companion galaxies to our own—represent brolgas, a species of bird. The large cloud is the male, the small the female. Native informants say that their color resembles that of the bird's egg. The birds breed between October and April, when the bright clouds ride high in the sky. Brolgas attract a lot of attention by performing an elaborate courtship dance consisting of lining up opposite one another, bowing, and bobbing their heads up and down as they advance and retreat.

Star and cloud patterns, too, are part of a Boorong eco-zodiac not unlike that of the people of the Andes. The set of patterns correlates a position of significance of each asterism in the sky with an earthly counterpart usually associated with the cycle of subsistence; for example, the Southern Cross stands straight up during March and April, and is linked with the bunya tree in which possums breed at this season, *Ichingel*, the Emu (made up of the Coalsack and α and β Centauri) is horizontal at midnight, signifying the position occupied by the female when she lays her eggs (April and May). *Otchocut*, the Great Fish (made up of the stars of our constellation of Delphinus), disappears in the west at sunset when cod spawn (November) in the receding floodwaters of the Murray River. And, yes, there is a kangaroo. They call it *Purra* and it is made up of bright Capella (of our Auriga, the Charioteer) and surrounding stars. Purra hops into the pre-dawn sky in early August and disappears in the early evening in late February. Boorong hunters stalk it at the time of the great midsummer heat (January and February), when the parched animal comes in search of water.

The examples of patterns in the sky I have chosen to discuss demonstrate one attempt by both our ancestors and by contemporary people isolated from the West to create and express order in nature as a guide to human action. We disseminate our moral order, express our good and bad behavior, our legal tenets, our desires, hopes, and wishes through a variety of stories, cast in the medium of print and images in books, on the internet, and on television. The tale of Orion in the Carib or Lakota worlds suggests that little has changed. Only the medium is different. Theirs is a vast two-dimensional, ever-changing dome of sky that exhibits a kaleidoscope of stars that form patterns brought to life by song and recitation, or in some instances only quietly contemplated.

Any references to nature and the sky in our modern stories have melted away, lost in a distant world of metaphor: "We met once in a blue moon"; "I love your starry eyes"; "Time and again I asked ..."; "same time next year"; "tomorrow"; or "in an hour." Today, but for Orion and maybe the Big Dipper, our old constellations go unrecognized, and when glimpsed they are only objects of passing romantic interest: "What a beautiful night it is! ... Where's the moon?"

In this chapter we have learned that patterns in the sky once offered practical knowledge—information about the weather, when to plant, when to harvest or go to war. Star patterns offered visible signals to prompt a multitude of both seasonal and daily events. And, as we shall discover in Chapter 8, they also came to signify transcendent power.

Did a single culture spawn the idea of peopling the sky with imaginary inhabitants and then spread the word round the world? It is a tempting idea, but based on the evidence I think it is safer to believe that we continually invent and reinvent the cosmos *after ourselves.* People construct zodiacs as a means of finding the familiar in the mysterious—the cave, the cloud, the sky. Thus we lend our worldly attributes to the unseen forces of nature and make them manifest in heaven. People all over the world who needed to watch the course of the golden sun would naturally assign names to familiar marking points along the great luminary's cosmic roadway—points which the moon, silvery ruler of the night sky also regularly visited. To make sense, constellation names would require a seasonal ring—like the watery goats and fishes.

The pursuit of wandering planets along the great skyway developed into a highly refined art among many cultures which sought to discover the influence the sky exerted on all creatures in the sub-lunar world. We shall deal with the nature of that above-below relationship in the chapter "The Astrologer's Sky" (Chapter 9). Some cultures became so preoccupied that they honed the sky-based calendar to levels of precision well beyond merely practical motives. Their actions would turn out to have great impact on social and political history. Later we'll explore some of these obsessions with celestial precision, but first we need to get a sense of the wide variety of contacts with the sky experienced by diverse segments of hierarchically organized cultures—I have chosen the sailor, the hunter, the farmer, and the family and kin group. Only then can we understand how complex organized societies fostered such extraordinary specialization.

CHAPTER 3

The Sailor's Sky

And Trail-of-light the one men call by name,
The other Twister. By it on the deep
Achaians gather where to sail their ships,
Phoinikians to her fellow trust at sea.
Twister is clear and easy to perceive,
Shining with ample light when night begins;
Though small the other, 'tis for sailors better,
For in a smaller orbit all revolves:
By it Sidonians make the straightest course.[1]

3rd-century BC Roman poet Aratus

Being a navigator in the Pacific Islands of Oceania in the 18th and 19th centuries was as respectable a profession as that of a neurosurgeon in our culture today. As one 18th-century navigator who visited the islands remarked: "Geography, navigation, and astronomy are known only to a few."[2] Getting from one place to another in an environment made up of 99 percent water, on an ocean affected by shifting wind, swells, and currents, required carefully cultivated skills. Astronomy was a major component of those skills.

To find an expert on the stars look for a *tiaborau*, an early explorer of the Gilbert Islands of eastern Micronesia was advised. The *tiaborau* was both a perspicacious sky watcher and an experienced navigator, who located and memorized the shapes and positions of linear star-to-star constellations on an imaginary three-dimensional compass. He knew which particular chain of stars that appeared over the horizon throughout the night would match which island bearing.

As we shall learn later, some seafarers of Oceania learned to use the "stick chart," a device that expressed the local topography of the ocean. This combination of map and speed indicator made it easier to travel over the surface of moving water, a task far more difficult than judging direction and travel time on land. Training in a "stone canoe" (we still find a handful of them dotting the shores of otherwise deserted coral atolls in the Gilberts) was an indispensable part of a seafaring astronomer's education. Like a modern flight simulator, these fixed structures were carefully aligned with the linear constellations to establish the appropriate directional bearings. Once internalized, the indicated directions could be implemented in a real sea-going craft that would take the navigator to specific island destinations beyond the visible horizon.

The navigators of Micronesia, Polynesia, and Melanesia, chains of tiny islands that straddle the mid-Pacific, owed at least part of their success at getting around to a unique sky condition that occurs in tropical latitudes. Suppose observers in two different latitudes face east after evening twilight. As Fig. 16 shows, their latitudes determine the angle of the path from/to the horizon followed by the rising and setting stars. The bottom frame depicts what the sky looks like to an observer positioned in the tropics, close to the earth's equator. The tropical navigator sees the stars rise more or less straight up in the east and plunge straight downward in the west (if you look in the opposite direction). The top frame shows what a mid-northern latitude observer witnesses. For most of us, as well as our Greek and Babylonian ancestors who lived in mid-northern latitudes, stars rise and set along highly inclined nightly paths. This geographic dictate offered navigators on the small Pacific atoll of Arorae in the Gilbert Islands a huge advantage.

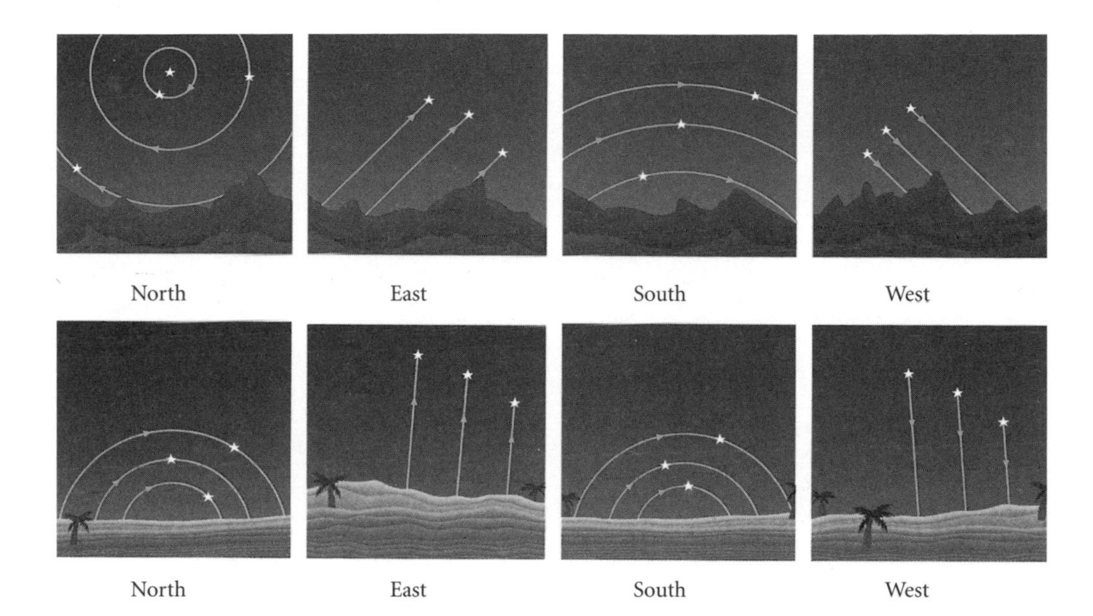

16 *What you see depends on where you live. Immediately above: looking in various directions along the horizon in tropical latitudes; Top: same view in mid-latitudes. For southern mid-latitudes star trails tilt in the opposite direction.*

Arorae's northern shore is dotted with half a dozen pairs of parallel rough-cut coral slabs, each about the size of a person. They are arranged horizontally and cemented into the ground (Fig. 17). One pair lines up with the neighboring island of Tamana, about 80 km (50 miles) distant; a second one targets Beru Island 140 km (87 miles) away, while a third points to distant Banaba at 700 km (435 miles), and so on. Islanders called them *Tetibu ni Borau*, or "the Stones for Voyages." Locals say these slabs were used as instructional models by their ancestors to set directions for inter-island navigation.

This is where astronomy enters the picture. Each pair of stones also aligns with the place where certain stars appear or disappear on the sea horizon at different times of the night. For example, at evening twilight in August the bright star Regulus aligns with the Tamana stone; at midnight Arcturus rises vertically to take its place. Both stars give the same bearing. The navigator simply memorizes a star path

17 *Carefully positioned stone slabs, called "stone canoes," were once used to teach navigational skills in parts of Micronesia. Here a pupil, seated at night between the stones, faces in one of the indicated directions and memorizes the linear constellations that rise in that direction throughout the night. In daylight the canoe's parts are used to illustrate sea swell interactions.*

(called *kavienga* in Tonga or *kavenga* in Tikopia) that consists of a long vertical chain of stars. The stars that make up the chain rise and set at approximately the same position, and their constant horizon bearing coincides with the direction to the island the navigator wishes to visit. He uses these linear constellations, or "star paths," to steer the canoe toward his island destination (Fig. 18). Star-rise positions are like points on a three-dimensional compass that can be memorized, and then transmitted orally from generation to generation.

Those of us who live in the far north (or south) would have difficulty making use of such a concept. For example, consider a North

Pacific or South Atlantic sailor who might try to adopt the system used at Arorae. As soon as his guide star appears over the waves, it would immediately begin to move laterally, away from the direction where it was initially sighted. Unless a substitute star is immediately available, the seafarer's course would deviate radically from a straight line. For example, sailing from New England to Great Britain, the course of a ship moving at 30 knots would be thrown off 10° or 8 km (5 miles) in an hour.

A modern navigator who once visited the local equivalent to his trade, the *tiaborau*, was shown a functioning model of a stone canoe located behind the family residence. Built by his father and based on an earlier version constructed by his grandfather, the pointed structure measured about 1.5 m (5 ft) east–west by 1.2 m (4 ft) north–south. It had a rectangular stone seat in the center and triangular rocks of different size and orientation represented the magnitudes and directions of ocean swells. A chunk of brain coral astride the seat represented the god of the sea. Curiously, the *tiaborau* referred to his

18 *A Polynesian seafarer navigates his outrigger through Pacific waters. He sets his course for a distant island, learned by memorizing linear constellations (in this case two of Orion's stars) in the stone canoe.*

stone canoe as "the island itself." At the time (the mid-1960s), the *tiaborau* still employed this simulator to instruct his daughter in the old ways of the navigator (Fig. 19). Unfortunately little evidence survives that can help us understand how this now largely abandoned navigational system developed.

Linear constellations and the sidereal compass, however, are not maps, in our sense of the word. They possess no analogues in the cosmology and astronomy of the civilizations of the Classical world that gave rise to our modern navigational systems. These ingenious tropical native techniques for navigating were prompted purely by environmental circumstances; they cleverly circumvent the conventional magnetic compass, sextant, and other astronomical contrivances of our own culture.

Investigators have recognized three-, five-, seven-, and nine-path navigational star systems in use over 5,000 km (3,100 miles) of the Pacific, from the Trobriands and Tuamotuans to Samoa. Unfortunately

19 *A navigator instructs his daughter in the use of a stone canoe.*

many of the indigenous star names provided by island informants were never taken seriously enough by outsiders to get transcribed onto western star charts; consequently we do not know the identity of most of the stars in their system.

On his South Pacific voyages Captain Cook had spoken of the way native navigators of the Pacific used bird-flight directions, swells and wind, and sun and stars to get about, but he never fully realized the extent to which the people he encountered had honed celestial navigation into a fine art. Much later, in 1890, a Tahitian navigator recited these instructions to a visitor for the first time:

> If you sail for Kahiki [Hawaii], you will discern new constellations and strange stars over the deep ocean. When you arrive at the Piko-o-Wakea [equator], you will lose sight of Hokupua [the Pole Star] and then Newe (?) will be the guiding star and the constellation of Humu (?) will stand as a guide above you.[3]

Unfortunately the names of the celestial guides in our language have been lost.

Ancient Hawaiian navigators, who lived just north of latitude 20°N, expressed their course settings in even more poetic language. In a system reminiscent of that of the Gilbert Islanders (we will meet them later in Chapter 6), they believed pillars in each of the four directions supported the sky. They named these pillars by their relation to the sun situated at the east point of the horizon (*alihilani*), along an east-to-west axis. North (*kukulu akau*) was the left-hand pillar (also meaning the direction "up"), and south (*kukulu hema*) was the right-hand pillar, or "down." The observer's position (*piko*) lay under the zenith point (*hikialoalo*), and the rising and setting points of celestial objects were called *hiki* and *kau*, respectively.

Another way to get around in the mid-Pacific was to use *fanakenga*, the stars the navigator recognized that passed the overhead position as seen from different islands. In our parlance, the declination (the angular distance north or south of the celestial equator) of a star that passes the zenith turns out to be the same as the latitude of the observer—another fact of geography. So every island in the mid-Pacific could be associated with one or more of its own zenithal guide stars. Again put in our terms, a navigator would know that he has reached the parallel of latitude of his destination when the arc of a guide star's course crossed the overhead position. (There is no evidence these people knew or cared about latitude as we know it.) Thus the Tongans say that Sirius is the *fanakenga*, or guide star, for the Fiji Islands (latitude 17°S); that is, it is "the star that points down to the Fijis." Likewise Rigel, in our constellation of Orion, is the *fanakenga* of the Solomon Islands (latitude 7°S), and Altair that of the Caroline Islands (latitude 9°N). Based on modern trial and error, navigators estimate that under calm conditions geographic latitude can be estimated to within ½° by sighting zenith stars. The double outrigger canoe used by native navigators provides surprising stability even on a rough ocean.

The Classical world also used its geographical idiosyncracies to create an efficient navigational system. Although at a disadvantage of being a considerable distance from the equator, the rectangular shape of the Mediterranean proved to be a plus, as it led in part to the invention of our system of latitude and longitude—the narrow dimension roughly lining up perpendicular to the axis of the globe. By the 6th century BC the Greeks had already conceived the earth to be a rotating sphere, and their love affair with geometry as the most pristine form of logic led them to impose a square grid upon the globe, the east–west direction being defined as longitude, the north–south as latitude, or distance from the equator.

Notwithstanding, as the 3rd-century BC Roman poet Aratus makes clear (see the epigraph to this chapter), ancient Night sends her signs to the navigator, and as anyone who glances at the *Iliad* or the *Odyssey* will discern, she has been sending them for quite some time. For example, skilled sailors knew which constellations rose at the same general bearing throughout the night. Not as precise as the point-to-point linear constellations of Oceania, the extended constellations nonetheless were used to give approximate compass directions. For example, once Odysseus had spread his sail following an encounter with Calypso:

> There he sat [at the steering oar] and never closed his eyes in sleep, but kept them on the Pleiades, or watched Boötes slowly set, or the Great Bear, nicknamed the Wain, which always wheels round in the same place and looks across at Orion the Hunter, with a wary eye. It was this constellation, the only one which never bathes in Ocean's Stream, that the wise goddess Calypso had told him to keep on his left hand as he made across the sea.[4]

Much later, Columbus employed celestial navigation (though with surprisingly little success) when crossing the Atlantic. To sail west at constant latitude he tried to keep Polaris at the same angle above the horizon off starboard. His principal tool was a calibrated quadrant, or clinometer, that measured star altitudes via the intersection of a 90° scale and the line of a suspended weight. But he seems to have misread the scales on his instrument and had trouble otherwise identifying the correct stars. Little wonder his early voyages meandered.

> The usual routine is to follow the star as it rises obliquely (or sets for that matter, if you are following a setting star) on the horizon. The rapid displacement of the star means you can only use the star roughly where the original one rose. Once the traveling-star has been displaced about 25° (roughly a handsbreadth at arms length) from its original spot it is far enough away to start looking for another star to follow. The star has to be very low on the horizon since your eyes are focused ahead of the dogs to look out for rough ice or any other problems in your path, and if you have to keep looking away and then back it can eventually become bothersome. You can be certain that if you have to raise your head to see the star then it is too high, and if you have to turn

> your head to see the star then it is too far to the side. Only low stars in sequence are any good for dog travel.[5]

This time the medium is snow, the vehicle a dogsled, the latitude 68°N (Coronation Gulf in northern Canada). And the navigator is an expert Inuit dog-team driver, who told an outsider that the whole purpose of star mythology was to teach his people to gain familiarity with the stars so they can be used for navigation. Each star and constellation has a legend behind it to pinpoint where it is. But the technique is much the same as that of the Pacific navigator, except that "when I am depending on a star to take me home I must not actually follow that star. I must follow behind the star, to the left."[6] Among the stars of the north used for Inuit navigation are *Quturjuuk* (Capella), *Sakiattiak* (the Pleaides), *Sivullik* (Arcturus), and of course the Big Dipper, or *Turktujuit*, which they identify with a caribou. When traveling over the ice and making for the shore of Baffin Island you raise your left hand toward the caribou. "Once your thumb and fingers match its stars your arm points toward the mainland."[7]

According to our way of thinking, navigating in latitude may be straightforward, but doing so in longitude is a different story. Because longitude is a measure dependent on the rotation of the earth, its accurate determination requires a precise timepiece on board. A difference of time between places basically amounts to a difference of longitude. Thus New York time is five hours (75°) earlier than London time. (360° or one trip around the world in longitude is equivalent to 24 hours of time measure; therefore one hour of time equals 15° of longitude.) It was not until the mid-18th century that ships' chronometers unaffected by movement on the sea were devised.

So much for our way of thinking. As far as we know, longitude (like latitude) was not conceptualized by indigenous sailors of the mid-Pacific—any more than their cosmologists dreamed up a Big Bang universe. There simply was no method known—much less any motive—for the islanders to deal with such a concept. Although Old World navigators found it necessary to use timekeeping mechanisms, ranging from sand clocks to chronometers, in order to chart the east–west segments of a course, the people of Oceania employed a knowledge of wind and oceanic currents combined with astronomical observation; that is, they relied on purely natural forces to get to where they wanted to go. That they managed to get about quite successfully proves that a knowledge of longitude and latitude are not indispensable concepts for all skilled navigators.

You're gliding in a canoe along a vast ocean at eye level with the rapidly moving water. You can see no mountains, no man-made structures, neither sighting posts nor permanent pathways to guide you on your way. The fixed sky is your solitary reference frame. Day by day as you slip past distant islands visible about the horizon, you begin to wonder: Who is really moving? Is my canoe fixed and are the islands the real objects in motion? While this may sound like a spinoff of the theory of relativity, such perceptions are the source of the system known as *etak*, used by the navigators of Puluwat Island in Micronesia to track distance interlinked with time on their voyages across the sea.

How does the *etak* system work? According to a modern navigator who spent years studying it,[8] the sailor selects an *etak* reference island for each particular voyage. Usually it lies 100 km (62 miles) lateral to the intended direction of the destination. Then star-path bearings are singled out from memory. In between, a navigator recognizes other reference stars that the island will pass beneath as it glides *backward* relative to the canoe. These mark out *etak* segments of the voyage. Just as you might chart out a flight from London to Cairo in segments based on successive countries your plane passes over, the navigator divides up the time-space of the whole trip into *etak* segments so that he can estimate what portion of the trip has been completed (Fig. 20).

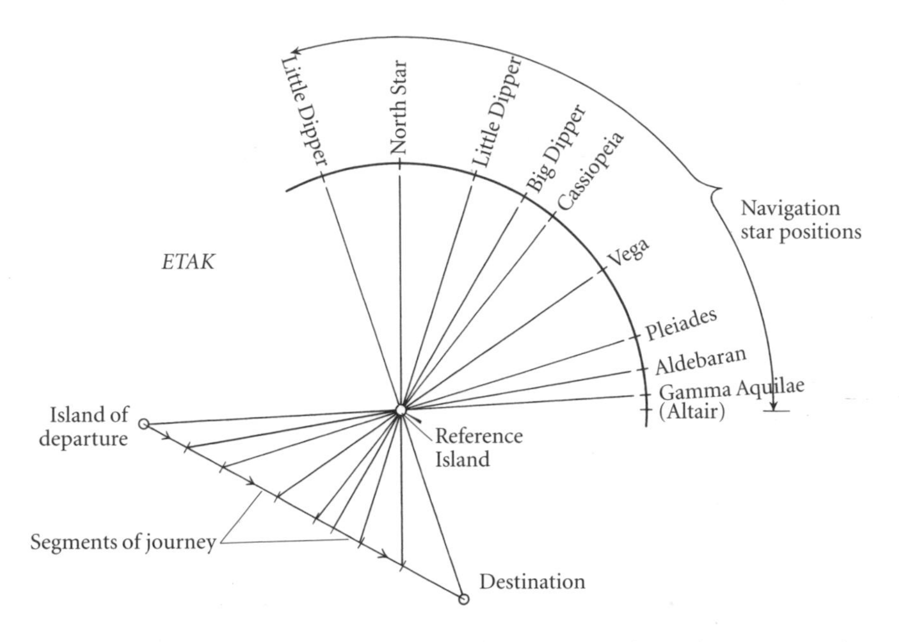

20 *Relativity at sea: the concept of* etak *divides one's journey according to the movement of a reference island with respect to background stars. By permission of Harvard University Press.*

A lengthy Tahitian sea chant collected by an early 19th-century preacher provides more clues. The following fragments give us an idea of how sea voyages were conceived in the mind of the native navigator. This particular passage contains explicit directions for setting a 4,000-km (2,500-mile) course from Tahiti to Hawaii. Note the way it evokes the *etak* concept by portraying an ocean horizon that vertically yields up images in a moving landscape:

Let more land grow from Hawaii
Mariua [Spica] is the star, Auere [Alpha Centauri] is the king
of Hawaii, the birthplace of lands.
The morning apparition rides
Upon the flying vapors of the chilly border.
Bear thou on! Bear on and strike where?
Strike upon Moana-urifa [a rank smelling
part of the sea], in the border of the west!
The sea casts up Vavau [Porapora Is.] the first born,

...
The sea casts up Arapa [island] alone;
Raparapa [island] alone
Just over the sea is Tai-Rio-aitu [Aldebaran].
Bear thou on! And swim where?
Swim towards the declining sun,
Swim towards Orion.
Distance will end at thine approach,
Redness will grow,
It will grow on the mountain figurehead
At thine approach,
Where the mountain is the boundary over there, O!
Angry flames shoot forth;
Redness grows, grows, grows on the figurehead
Bounding in
The ocean over there ...[9]

The idea of blending time and space in Polynesian ocean travel is further reflected in another exotic-looking device once used in the Marshall Islands, located in the western Pacific about 4,000 km (2,500 miles) southwest of Hawaii. First described by western navigators in the 19th century, the "stick chart" is a complex reference scheme related to the *etak* concept of space–time and reminiscent of the stone canoe. Only a few dozen authentic stick charts survive, but, thanks to some early ethnological accounts, we know a good deal about how they work. There are three kinds: instructional charts for training, localized charts, and general charts that cover large areas. A typical stick chart consists of a network of straight and curved reeds bound together to form a flat frame (generally about 75 × 75 cm (30 × 30 in.)) that can be held conveniently in the hand (Fig. 21).

In one representation, the seafarer thinks of the stick chart as a map that tells him where to proceed. The straight and curved lines represent the course, while certain of their crossing points signify the location of the islands. These are often symbolized by conch shells

21 *A network of straight and curved reeds bound together form a "stick chart," a navigational device used by the Marshall Islanders of Micronesia as an aid to memorizing wave and swell patterns that occur between their islands. The central vertical sticks represent the course to be followed, while the bent reeds on each side signify the eastward- and westward-moving swells refracted by the island. The Marshalls, a double chain of parallel islands, are especially noteworthy for the interference patterns they produce in oceanic swells. This is the likely place where the stick chart was invented.*

fastened to the reed matrix. But the navigator also can use the curves and the intersections on the chart to call to memory the behavior of wave patterns experienced on the ocean's surface.

What further information do the stick charts contain that helps the navigator find his way? First, there are swell patterns. Have you ever noticed that waves that pass a protruding object, say a rock jutting up out of a shallow lake, seem to bend around that object as they pass it? Likewise, when a large wind-guided swell approaches an island its ridge pattern gets modified. It curves like a parenthesis, as if to embrace the island, looking more pronounced the closer it gets to the shore. Once the swell reaches the island it dissolves into smaller waves that finally break upon the reefs. An experienced navigator can detect good-sized swells well out of sight of landfall, not by his eye but by his body; for example, he can sense delicate swaying motions or feel slight differences in the pressure of the mast upon his thigh.

A second type of disturbance is the counter-swell; it happens on the leeward side of the island and it too seems to embrace the land as it nears it. But what if a navigator failed to recognize the curving swell and sailed past his destination? He could detect his error by recognizing the pattern of disturbance along the line of nodes where the opposing swells meet. He could sail along this line in the direction of increasing turbulence until he sighted the island. In Fig. 21 the vertical axis of the stick chart represents such a course, and the bent reeds symmetric about the axis signify the eastward- and westward-moving swells. In this case the island would lie at the center of the chart.

It is important to realize that the two interpretations of the stick chart, both as a map and as a taxonomy of disturbance patterns, were often engaged simultaneously. Like *etak*, stick-chart navigation is a very difficult notion to fathom for those of us who are used to separating time and distance. In addition to mapping each of these disturbances onto his highly personalized stick chart, the navigator could also indicate the crossings of swells of one island with another. These observations of ocean wave interference patterns, many of them undetectable by oceanographers until the advent of satellite imagery, complemented by mental notations of wind direction, sea bird sightings, the presence of luminescent organisms on the ocean surface, and of course sightings of the sun and the stars, were all part of the skill set of the exalted tropical navigator.

And so, the environment of earth and sky provides an anchor for fixing the direction of human thought. Like the invention of linear constellations, which take advantage of the peculiarities of sky motion in the tropics, the stick chart was created as a response to the way nature behaves in the local environment, which, for most people, is the only one that matters. In this case the Marshall Islands constitute a navigational corridor produced by two chains of islands that run 50 to 100 km (30 to 60 miles) apart for nearly 1,000 km (620 miles). Many different kinds of wave interference patterns occur within this region. Because navigation was the most honored profession—and most valued activity—in the cultures that once populated the tiny atolls of the mid-Pacific, navigators needed to develop a means of recording and communicating practical knowledge about the ocean environment. Not surprisingly, the material elements involved in the expression of this knowledge consisted of reeds, shells, and coral slabs, the most readily available products on the island shores. By using their ingenious invention, Marshallese navigators transformed sensory impressions of their watery environment into concrete mnemonic schemes that helped them deal with the central problem of navigation—getting there.

Evolutionary biologist and writer Jared Diamond has characterized the prehistoric Polynesian expansion as "the most dramatic overwater exploration in human history."[10] One of the great questions about the cultures of Oceania is: How did they get there from the mainland in the first place? In many instances hundreds of miles separate segments of the great island chain. We are offered a clue by the fact that much of this area comprises low-latitude locations, providing ideal environmental circumstances for linear constellations and a sidereal compass. Our knowledge of their navigational techniques also helps answer the question how. Deep-sea voyaging across the southeast Asian archipelago may seem very difficult, but modern anthropologists and adventurers have used native technical skills to reconstruct and carry out such voyages themselves, so there is little question that it *could* have been done.[11]

Why sail to the islands in the first place? There would have been no shortage of practical motives beyond pure adventure for conducting long-distance trips across vast expanses of water: to trade material goods, to conduct raids or warfare, to extend a chiefdom's sovereignty, and to obtain food and other items from uninhabited islands (one group of Cook Islanders was said to have traveled over 300 km (185 miles) just to procure supplies of birds' eggs). Economic disruption at any time on the mainland would have offered a logical motive to force a colonial type of expansion out into the more sparsely inhabited or totally uninhabited islands even farther off the coast. (It is well known that the city of Venice in Italy was founded shortly after the fall of the Roman empire in the 5th century AD. Invaders from the north began to intrude upon the relatively stable resource base of the northern Latin cultures, which for centuries held sway over the rich fishing grounds in the lagoon.) Indeed, celestial navigation may have become the most honored profession among the civilized people of Oceania through a cultural adaptation of mainland people, necessary for their survival in a future water-based location.

Today we travel by car or we entrust our itinerary over the planet to bus drivers, train engineers, or plane pilots. Thanks to technology, none of them have any need of looking up. Nor do we. Only the handful of us who navigate for pleasure are aware of how important the sky once was for setting direction. But as we shall see in the next chapters, navigation was not the only valued star-dependent profession.

CHAPTER 4

The Hunter's Sky

I don't go far in the beginning. I go some distance and come back again, then, in another direction, and come back, and then again in another direction. Gradually I know how everything is, and then I can go out far without losing my way.[1]

An Australian Aborigine

How do you keep track of time and how do you reckon your position in space if you're constantly on the move? Those of us who spend a lot of time on the road have clocks, calendars, and maps to guide us as we move from place to place; but those without sophisticated technology to mechanically replicate the workings of nature, and who had no permanent place they called home, can acquire information about time and place only by confronting nature directly.

There are many ways to do this if you let nature be your guide. For example, you can watch birds, clouds; observe trees, ant hills, sand dunes—even consult the wind direction. Different species of migratory birds each have their own flight patterns. (The migratory path of the Polynesians in their colonization of the Pacific, which began about 1500 BC, follows one such route.) Native Arctic migrants are skilled at detecting large open bodies of water at great distances by observing clouds—the telltale "water sky" casts dark patches on their undersides. Some clouds—standing clouds—remain fixed over the high peaks they envelop and thus locate the mountains for the observer. The Iroquois of the northeast U.S. knew the St Lawrence River was not far off when, especially on warm days following cool nights or a sudden rainstorm, they spotted distant clouds, which marked the position of small islands in the estuary. Driving home from the east on wintry days I have frequently noticed, even at distances up to 150 km (100 miles), the increased cloudiness over the southeastern shore of Lake Ontario that produces "lake effect" snow (bands of intense snowfall downwind from the lake). Wind can also be an indicator—prevailing winds (those that blow longest from one direction) leave their mark on trees, retarding growth on the side from which they blow. The axial directions of large crescent-shaped dunes also indicate wind direction.

In 1724, Father Joseph Lafitau, who lived among the Iroquois for five years, wrote of their tree compass. Tree tops, he said:

> always lean towards the South, to which they are attracted by the sun … their bark, which is more dull and dark on the north side … [and] the various tree rings which are formed on the trunk of the tree are thicker on the side which faces north than on the south side.[2]

In cool climates of northern Europe—and especially in open areas—ants build their hills with a higher peak on the southeastern side to acquire the greatest amount of warmth from the early morning sun. As a result, the hill slopes away on the northeastern side, forming, in effect, a directional compass. In the Australian desert termites build their elongated mounds, some of them up to 4.5 m (15 ft) high, so that they align exactly north–south. This is because they are built under wet conditions, and they dry most efficiently if the broad sides face in the direction of the morning (eastern) and afternoon (western) sun.

An experienced nose is quite capable of sensing game at great distances. While walking recently I saw a herd of deer prance across the road some 150 m (500 ft) in front of me. Well before I reached the spot

I was able to detect their scent quite clearly even though they had moved well beyond my view. Bring along the even more sensitive apparatus of the hunter's four-legged companion, whose nose lies closer to the ground where the scents are strongest, and you've increased your chances of bagging a nourishing supper.

All these factors (see Fig. 22 for some examples) can play a role in trail making. The learning process begins with a careful eye accompanied by repetitive observation and exploration by successive approximation, as my first epigraph suggests. Once a trail is learned, it can be marked permanently. All scouts know the technique of cutting marks in the trunks of trees at select intervals. Some Native American tribes would stretch a branch away from a young tree and bury it along the direction of the path. There it would take root and grow in that fixed position making a permanent road sign.

The sun, moon, and stars are easily the most reliable aids to nonsedentary pathfinders. Set a course true north by finding the pointer stars (the two stars that form the right hand side of the 'bowl') of the Big Dipper, and following them to a distance of six times the interval between them. That gets you to the north celestial pole, conveniently marked by Polaris, the Pole Star. Walk in the direction straight below it and you'll travel due north. If you keep the pole star a fixed angular distance above the horizon as you move (this is easily measurable with spans of fingers held horizontally at arms length), then you'll be moving east (if the pole star is to your left) or west (if it is to your right). Though there is no bright star to mark it, southern hemisphere observers can locate the south celestial pole by bisecting a line between bright Achernar and a point midway between the Southern Cross and α and β Centauri.

Sun directions are readily available at noon when the shadow of a vertical stick is shortest. To get true east–west, draw a circle on the ground. Place a vertical stick at the center and mark the points where the shadow tip crosses the circle. (The shadow crosses the circle twice,

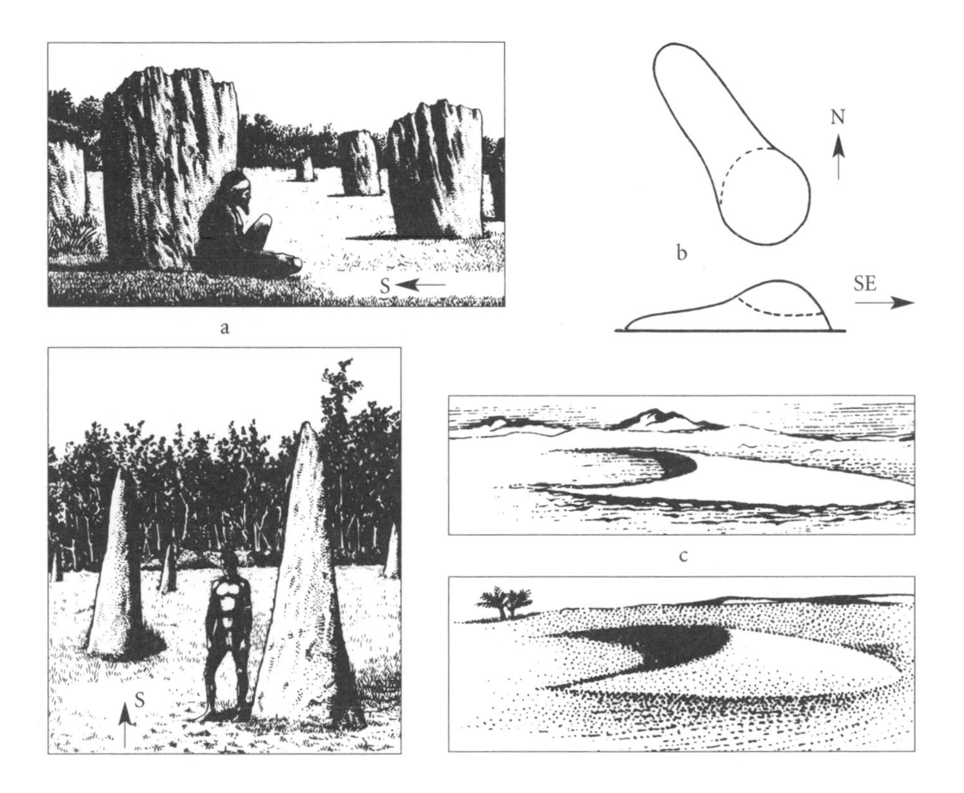

22 *Let nature be your guide:* **a** *two views of termite mounds,* **b** *compass anthills, and* **c** *barkhan (dunes) are just a few examples used in direction-finding around the world.*

once in the morning and once in the afternoon.) The line connecting the points marks the east–west direction. The moon can be even more convenient for pinning down both date and time, as it changes more rapidly. For example, catch the waning last quarter moon rising and you will know it is approximately midnight.

Suppose it is the beginning of spring and you want to plan a night-time summer trip of ten days that optimizes the benefit of the light of the moon. Note the date of the first full moon after the spring equinox. Back up one day per month to get the date of the next full moon. Thus if the moon is full on 29 March, consecutive full moons will also fall on or about 28 April, 27 May, 26 June, 25 July, and 24 August. (In the real world of the indigenous calendar keeper all of the

The Big Dipper Clock

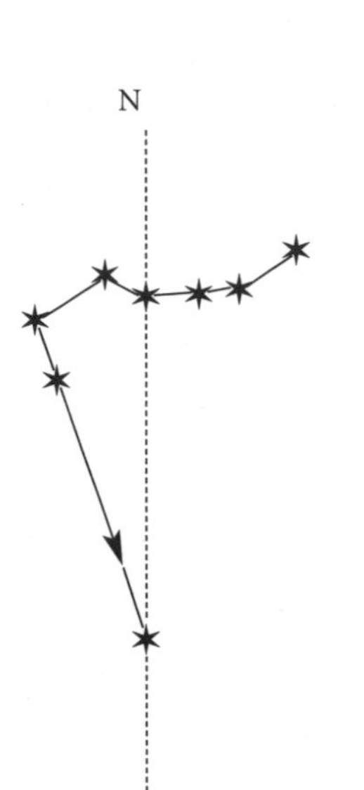

You can use the pointer stars of the Big Dipper (the two stars at the end of the bowl) as the hour hand of a clock, with Polaris as the minute hand. For example, if the pointers line up the eleven o'clock position from Polaris, as shown in the drawing, subtract eleven from twelve, multiply by two and add eleven:

$$12 - 11 = 1$$
$$1 \times 2 = 2$$
$$2 + 11 = 13$$

From this number subtract two hours per month from the last spring equinox (21 March) and 4 minutes for each additional day. Thus, if the date is 25 July, four months and four days from 21 March, subtract 2 × 4 = 8 hours, and 4 × 4 = 16 minutes. Thus

13 hours – 8 hours 16 minutes = 4 hours 44 minutes, or 4:44 a.m.

(In the modern world at certain times of the year you would need to add an hour for Daylight Saving Time, where applicable.) Needless to say a hypothetical hunter-gatherer would have internalized such a process, committing to memory the relation between Dipper, orientation, date, and interval remaining until morning twilight. The arithmetic would be eliminated.

"higher math" in my hypothetical exercises would be done on the fingers.) Since you know the rains are heaviest in late July and August, the best time to travel might be the ten days beginning on 26 June, when the waning moon lights most of the night sky.

Successive full moons are often named and correlated with the human activities that accompany them. Most familiar are those we derive from the native tribes of the northeast U.S. The Harvest Moon is the one that occurs closest to the autumn equinox. It is followed by the Hunter's Moon, when the deer are fattest. Then comes Beaver Moon which is associated with trap setting, then Cold Moon, Wolf Moon, and Snow Moon when little activity takes place. Once the ground begins to heat up comes Worm Moon (the worms come out), and so on. Indigenous month names can betray both climate and culture. Thus, Turnaround Moon, Cythere Flower Moon, Bulls Seek Shade Moon, Guinea Fowls Sleep Moon, and Rains Rot the Ropes (used to tie cattle) Moon are months named by the coastal people of Madagascar, while the Siberian Ostiak reckon the Ducks and Geese Go Away Moon, Pine Sapwood Moon, and the Moon In Which Men Go On Horseback. Consecutively named Inuit months beginning with spring equinox are Sun Is Possible, Sun Gets Higher, Premature Seal Pups, Seal Pups, Bearded Seal Pups, Caribou Calves, and so on. In Swazi (South Africa) time these same periods are called Small Elephant, Large Elephant, Fire Making, Brilliant Star, Aloe, and Black-Shouldered Kite.

We cannot overestimate the value of the lost skills, many of which I've just outlined, that our hunter-gatherer ancestors employed for obtaining precise knowledge by carefully and persistently observing the detailed behavior of the world around them. Western contemporary culture, so deeply immersed in a technological way of life, tends to underestimate the ability of our predecessors to acquire precise knowledge from the observation of nature. I think this is part of the reason why we often conjure up ancient alien astronauts as a source of the handiwork of native people who once lived very close to the earth.

Hunter-gatherers have long been stereotyped as "primitive people" who randomly shamble about the landscape in search of sustenance.

Nonetheless, like sedentary people who successfully farm, build villages, and erect cities, the people on the move who thrive are those who are best organized. Just as the farmer needs to know when to plant (see Chapter 5), the gatherer benefits from environmental cues as to where and in what order particular wild forms of vegetation ripen. For example, the Mallee aborigines of northeast Australia know that when Arcturus appears in the northeast in early evening (early March in our calendar), wood-ant larvae are ready to be collected. Likewise, the hunter requires a knowledge of the migratory routes and schedules followed by the game needed to sustain the extended family for the lengthy winter ahead. When Arcturus makes its morning heliacal rise it is time for the Mallee to move to the sea near the plains, to hunt the magpie geese who flock to feed on newly sprouted water chestnuts.

How long can we remain on this side of the river before the flood season will hinder crossing back? When can we anticipate finding the birds' eggs we once discovered here? Will there be significant time to get our supplies into winter storage before we need to move on to the hunting ground? An intimate knowledge of nature's signs demands both a good eye and a means of remembering and passing on collected information. Recognizable changes in nature follow the rhythm of the sun and can easily be correlated with the place on the horizon where the sun rises or sets, as seen from a fixed location that can be visited periodically—call it a "sun station." To calculate short duration periods between events one could count days on the fingers and pass on the knowledge orally—or indicate it by tally marks carved on wood or bone, or engraved on stone.

As we have seen, a visible short-term-cycle clock with a recognizably changing face moves across all our skies. The moon illuminates the sky as it passes through its 29½-day cycle of phases. There is good evidence that hunter-gatherers employed this natural clock as early as 30,000 years ago.

Recovered from a cave in central France, the artifact shown in Plate 1 consists of a carved piece of bone that was once part of an eagle's wing. Microscopic examination of the artifact reveals that by holding the bone in the palm of one hand, the user must have applied a twisting motion to gouge out each mark with a stylus or pointed tool. Upon reaching one end of the bone, he/she turned it through 180° in the hand and created the linear pattern that runs in the opposite direction. The twist of the comma-shaped marks visible under the microscope proves that they are ordered sequentially. But for what purpose? What had previously been thought to be the marks of a tool sharpener is more likely a notational record of the days of the lunar phase cycle, each point standing for one day. Note the single continuous, sinuous pattern (shown in the diagram of Plate 1) that links the dots. These gouged-out points can be connected by a single sweeping curve—a boustrophedon-type notation. Now count and group the dots, and you'll notice that each line consists of clumps of fifteen (give or take one or two points). These intervals correspond to major changes in the lunar phase cycle. The bone carver may have been the first person in the world to keep a record of what transpired in the heavens. But why would a hypothetical Paleolithic calendar keeper have created such a lunar record? What was its purpose?

A 19th-century Ojibwa woman from Canada cites a modern example of this ancient practice in her diary that may offer a clue:

> My father kept count of the days on a stick. He had a stick long enough to last a year and he always began a new stick in the fall. He cut a big notch for the first day of the new moon and a small notch for each of the other days. I will begin my story at the time he began counting a new stick.[3]

She then goes on to describe how her mother began storing goods, such as maple sugar and rice, for the winter during the first moon phase interval.

The Changing Face of the Moon

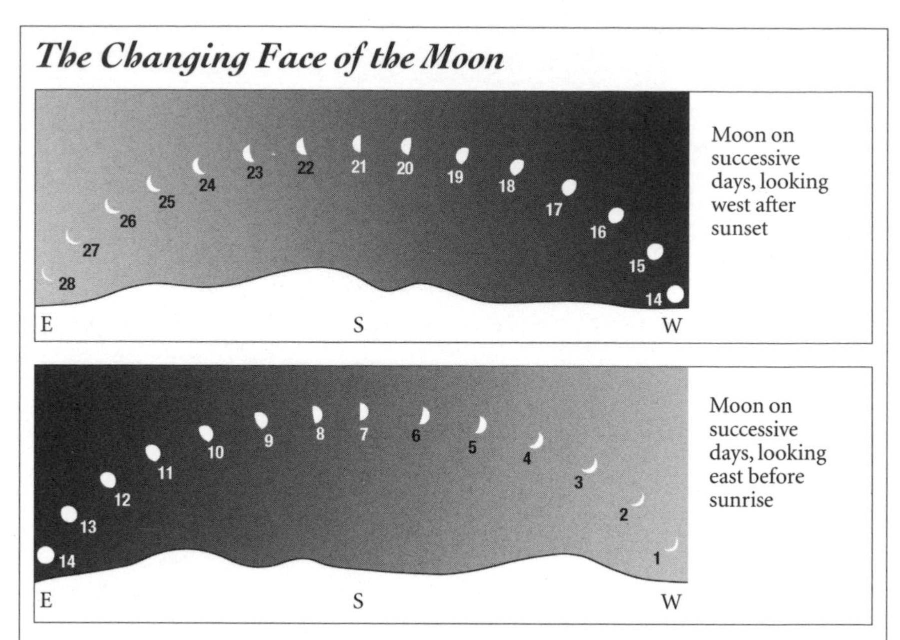

Phases of the Moon showing positions relative to the sun.

The fastest mover along the zodiac, completing its cycle in a month, the moon, like sun and planets, treks eastward among the stars from night to night. But the moon has one property that is unique to it—it changes its face. It begins its cycle with first visibility in the sky as a thin crescent, its horns pointing away from the setting sun, over which it can be briefly sighted (bottom right). The next night the moon appears at dusk slightly higher in the sky, and higher still on the third night as it waxes to a quarter, then it grows to gibbous (when more than half the disk is visible), then full phase. By that time, fourteen or fifteen days after it has first been sighted, the full moon rises in the east directly opposite the setting sun, and it sets directly in the west opposite the rising sun (bottom left and top right), the observer appearing to stand on a direct line between the two in space.

Once the half-cycle of phases is complete, the moon proceeds to wane from gibbous back through quarter—now more visible to the daytime observer—until finally only a disappearing thin crescent, positioned just above the rising sun in the east, remains (top left). Then the cycle begins anew.

Imagine a Paleolithic moonwatcher sitting by the fire at the mouth of a cave in early evening twilight. He notices the first thin crescent moon low in the west. He makes a mark on the wall of the cave or on a piece of bone held in his hand. Next night he sees a thicker crescent slightly higher in the sky; he makes a second mark, and so on. By the time the sixth or seventh mark is made, a D-shaped quarter moon stands high on the meridian at sunset. He notices that by this point in the cycle the moon has begun to provide a substantial amount of faint light for a few hours after sunset. The gibbous phase that follows brings with it a yet longer duration of even brighter moonlight. This signals to the hunter that he will be allowed more time for activity after dark, such as fishing or hunting nocturnal animals. (See Box: *The Changing Face of the Moon.*)

Or perhaps our ancient moonwatcher is a woman who recognizes that her menstrual cycle commences with that first visible crescent. We will never know. For man or woman the moon can be a reliable standard for mapping out cycles of activities and for telling and retelling the stories that embrace them. It offers clues to what anthropologists call an "adaptive advantage." Your eye on the sky gives you a precious glimpse into the future.

Archaic petroglyphs, carved by semi-nomadic peoples on the desert periphery of the Great Basin of North America, offer further clues about how ancient hunter-gatherers took advantage of the sky. Abstract figures such as meanders (are they water or snake symbols?) and spirals (perhaps a journey, or the turning around of the year?) dot the walls of canyons up and down the usually parched riverbeds of Mexico's northern state of Nuevo León. Some signs are rather more concrete; for example, the head of a reindeer with a score of dots positioned between the antlers likely represents a kill tally, while a tree in profile with adjacent strike marks might indicate the size of a gathering.

These vivid testimonies in stone, wood, and bone sometimes include markings that offer evidence of the possible use of the moon to schedule the activities of people on the move. Take the 1 × 2 m (3 ft 2 in × 6 ft 6 in) stone slab on the hilltop site of Presa de la Mula, located on the southeastern edge of the Great Basin in Nuevo León (Plate 2). It exhibits 207 vertical strike marks suspended from six horizontal lines, each compartmentalized into vertical cells by three straight lines. (At Boca de Potrerillos, another petroglyphic site 40 km (25 miles) to the east, a second dot pattern yields exactly the same total.) Seven lunar months total 207 days, not an unreasonable period to record if you are a member of a society never fixed in one place for a full seasonal cycle. Perhaps they spent only the hot months, April through October, there. Winter in mountainous northern Mexico can be quite harsh.

Examined close up, the groups of markings reveal that certain combinations of numbers repeat themselves. When added together and set off from one another by small circular "completion marks" they appear to record the number of days in a cycle of phases of the moon. One cluster counts the number of days from first to last quarter, tacking on the last five days until the disappearance of the last crescent. Another splits the cycle in two, measuring the interval from first crescent to full and then from full to last crescent. Still a third array segments the phases into quarters. The way the count is presented on the stone in slightly unequal intervals has led to the conclusion that these early moonwatchers were reckoning visible moon days exactly as they saw them,[4] for indeed when measured by the true extent of lunar visibility, some months really are longer than others.

Lunar changes were obvious harbingers of things to come; lunar scheduling provided not only an adaptive advantage for the ancient hunter-gatherer, but also signaled a measure of order amid events that otherwise might seem random or chaotic. Contemporary hunter-gatherers also use a similar strategy.

At the beginning of the 20th century anthropologists believed primitive man represented an earlier stage of evolution, one that found him closer to his origins. According to this theory all civilizations evolve from hunter-gatherer to chiefdom to state, climbing a ladder of progress toward more complex hierarchical organization and specialization. While the archaeological record substantiates the evolution of complex societies out of simpler ones, I believe that the introduction of the word "progress" into the evolutionary scheme can be misleading. The implication of the theory of progress is that whatever kind of astronomy was practiced by humanity's most distant "ancestors," is now long gone from the planet (except in rare cases of people who live in total isolation). Such ideas about cultural evolution were based on grossly misapplied notions of complexity derived from Darwin's theory of the evolution of biological organisms. They have since been largely discredited. Anthropologists find that skywatching among contemporary hunter-gatherer societies, though dramatically affected by outside contact, exhibits many of the same characteristics left in the archaeological record by prehistoric people.

African people who hunted and gathered in historical times were long referred to by the term "bushmen," an erroneous label that not only eliminates the female gender but also implies no identity or social common bond, no formal language, not even a physical type.

When anthropologists first studied them in the 1960s, the G/wi (the / is a dental click, as in our "tsk-tsk") people numbered about 3,000. They moved about in a 130,000-square-km (50,000-square-mile) area in northern Botswana that spanned a swamp, a huge lake, and a desert—the central Kalahari. The G/wi trapped giraffe, hunted antelope and wildebeest, tortoises, and termites; they collected roots, tubers, and berries; they also tanned leather, traded furs, did handcraft woodwork, and domesticated oxen. They accomplished all these tasks in a diverse, harsh environment conditioned by heavy rain, drought, frost, and searing heat. They housed themselves in temporary dwellings in at least three different encampments across five named seasons:

!hosa "hot time when trees flower" (which corresponds to our August, the end of winter, to December, when first rains begin)
N//aosa "raintime when antelope breed and grass is green" (! and // are other varietal click sounds) (December to early April)
Badasa "when Tsama melons are plentiful, rains cease and the veld dries out" (early April to mid-May)
G!wabasa "when plants begin to die" (mid-May to mid-June)
Saosa "cold time" (mid-June to August)

Some G/wi star names reflect a connection with the seasons via a correlation between their first appearance and what happens on earth at that time of year. Thus the bright stars of the Southern Cross are the eyes of giraffes; Orion (again a hunter; see Chapter 2) is a man shooting a wildebeest; and Regulus is the firewood finisher because it sets when the firewood supply is gone. The G/wi used the phases of the moon in a similar fashion. "I found it convenient to refer to the phases of the moon to indicate to informants the timing of my own past and future movements ..."[5] wrote one anthropologist who lived among them for a period of eight years.

The Julawasi !Kung, who roam the driest part of the Kalahari, carefully time the rainy season. They call the bright star Capella the "green leaf horn" because these flowers predict the onset of the rainy season—they bloom just a few days before the first rainfall, when Capella reappears in the evening sky. It also marks the time when the !Kung break up into small groups. They also call the stars of Orion's belt "the three zebras" because their appearance marks the commencement of the bow-and-arrow hunting season.

Among the North American Arctic Inuit the Aagjuuk stars make up the most important constellations. Aagjuuk consists of Altair, in our constellation of Aquila the Eagle, and the two fainter stars, β and γ Aquilae, that flank it. When first seen in the morning twilight they are thought to bring the light of the new year. In fact, the name derives from the arrow that refers to the upward pointing beam of sunlight cast upon Aagjuuk's return above the horizon; for these Arctic Circle dwellers this happens at the end of December. But the Aagjuuk stars also signal the time when the bearded seals begin their migration from the open sea toward the ice-encrusted shore and consequently can be hunted, though not before a great celebratory feast is held in order to give strength to the land. The Aagjuuk is recognized as far east as eastern Greenland, where natives use the first appearance of Altair to set up their annual calendar. The year begins with the first visible moon following Altair's official observation. The assurance of the return of light and life to this harsh environment is ordained in the stars.

Accumulated time seems to mean little to most hunter-gatherer societies. Long-term historical chronologies, and the idea of adding up small cycles to make bigger ones, except for the accumulation of months to make a year, simply is not as important to them as it is to us.

For example, if you ask an African Mursi tribesman what *bergu*, or month, it is, he might tell you that some people in his village recently told him it was the fifth while others said that it was the sixth. Disagreements about what time it is can become as heated as a close call at home plate in a baseball game. This built-in disagreement about time was once taken by anthropologists to imply that these semi-nomadic Ethiopians, who live by and depend upon the ecology of the Omo River north of Lake Turkana (Lake Rudolf), do not care about time, or at least cannot reckon it. What the anthropologists failed to realize is that for Mursi people, who are on the move about half the seasonal year, a calendar is not the same as the thin booklet of paper we hang on our wall. For them, keeping time is an interactive process, a dialogue among many people based on social rules that advocate an agreement to disagree.

What motivates the need to argue and disagree about time? To answer this question we need to analyze their system briefly. The

Mursi count twelve full moons in a season, and every tribesman can recite the set of social and agricultural activities that go with the lunations, or *bergu*. For example, in *bergu* number one people select and move into riverside cultivation sites; in *bergu* two they clear the vegetation; in *bergu* three they plant sorghum, which they harvest in *bergu* five. By *bergu* seven they are preparing the surrounding bushland for cultivation. The eighth lunation is when the great transition from dry to wet seasons usually occurs, for that is when the lengthy rains begin. After harvesting the crop in *bergu* eleven, they start to gather honey. The twelfth moon is one of celebrating year's end, a period of feasting and beer drinking. Now twelve moons add up to 354 days, about eleven days short of a seasonal year measured by the return of the sun to the same place on the local horizon. The heated debate centers on the Mursi way of responding to the imperfect fit between lunations and seasons, to which the Romans in Caesar's time reacted by striking moon-time out of existence and substituting sun-time in its place (see Chapter 10). But unlike the Romans, the Mursi had no interest in keeping long-term records.

Following the last *bergu*, Mursi timekeepers created the period they call *gamwe*, an interval of activity that lay outside the calendar. Calendrical time-outs are common, like the extra five days at the end of the year in the Egyptian or Maya calendar. Another example stems from our familiar twelve days of Christmas, the leftover handful of days by which the seasonal year exceeds twelve lunations, a product of the old pagan calendar. (On a shorter time scale, many of us think of our weekends as time-outs from the normal course of affairs.)

The Mursi *gamwe* measures a full lunation, resulting in a thirteen-month year that exceeds the seasonal year by about eighteen days, rather than a twelve-month cycle that falls eleven days short. But how to recycle the year to make things fit? We might add, or intercalate, a month every few years to synchronize the solar and lunar cycles. But if our sky cycles were tied to other natural periods—say a water cycle—

we might not be quite so disposed. Indeed the *gamwe* is more subjectively determined than the other *bergu*. That's because it is tied to the flooding of the river and not to the more objectively classified phases of the moon. The river will always crest in *gamwe*; how high it crests will determine how extensive the flood will be. Though the Mursi reckon a *gamwe* in every *bergu* cycle, the floods occur neither at the same time nor to the same extent in all Mursi land. What one person sees, another may view quite differently. This answers the question of why the Mursi argue and disagree about time. Unless people are willing to haggle over the calendar—even changing their opinion about what time it is—they will not be able to integrate all their valuable observations of nature, which helps the cohesion of the wider community. Clearly, skywatching is not separable from social concerns.

Mursi flex time is well informed by astronomical indications that fuel the fires of debate. For example, the names of the constellations, like those of the *bergu*, are tied directly to the subsistence activities on which they so vitally depend. Thus, the last appearance of the Pleiades in the west happens in *bergu* nine, and this star group is overhead at sunset in *bergu* six. When Sirius and Canopus appear together in early evening the Mursi know they must return the cattle to the homestead from grazing, for then it is *bergu* five. The Mursi also monitor the flooding of the river by the stars. When the right-hand star of the Southern Cross (called *Imai*) disappears in the glow of evening twilight, they say that the river will have risen high enough to flatten out the imai grass that thrives along its banks. By the time β Centauri (*waar*) disappears, the Omo River, which the Mursi also call Waar, has risen to its maximum level and overflowed into the low forest. The *sholbi* tree flowers along the Omo's banks when the star Sholbi (α Centauri) disappears from the sky. These astral rules of thumb are a reminder of many skilled observations made by the Mursi's predecessors, who farmed and raised cattle well outside the sphere of colonialism. For the Mursi, time is activity itself, not just some abstract measurement reckoned on a clock.

One of the most commonly held views among people who live in close contact with nature is that the gods are our ancestors and that they once lived among us. As we remember from Chapter 1, creation stories often suggest that, having made the world fit for people, the gods retreated to heaven, which had become separated from the earth. There they live today, and we must look to them when we pray in order to pay them our debt. The gods speak to us via omens that can be interpreted by the wisest in the group—those appointed to carefully watch events unfold in the sky. For the sedentary who reside in the city (see Chapter 7), the channel of communication is the temple, carefully arranged on sacred ground. But for the hunter-gatherers, who are organized in small families and clans, temples are modest affairs. Simple sun-watching stations can be lodges, grottoes, hilltops, bends in rivers, artesian wells, or special places at the edge of the sea, whatever offers an awe-inspiring access to nature—and especially to the sky.

As people band together in semi-sedentary societies, tending cattle, planting crops and perhaps irrigating the land during one season of the year, and herding or going on extended hunts during another, their efficiency yields them a measure of time to embellish their more permanent place of habitation by constructing temples. They might cut steps and build altars in the sacred cave or on the sacred rock; or they might construct a house of the sun, perhaps a small pyramid to imitate their sacred mountain. Degrees of separation occasioned by increased specialization that comes with more permanent habitation relegates the intricate detail of ritual to a priestly class close to the family patriarch, who himself often functions as both the wise and holy one. If it serves their mutual benefit, extended families join together to form chiefdoms. The place of worship becomes still larger and more elaborate, for now it must serve the needs of a group that may number in the hundreds. Clans meet in the inner sanctum of this sacred space to bond, not just by worshipping the same gods in the same manner, but also by sharing food and trading goods.

One of the most majestic sacred places whose layout and arrangement was closely patterned after the sky survives today in southern Great Britain. It is one of the ancient world's most famous monuments. Our brief visit to it will dispel all doubts that pre-literate, semi-sedentary people were incapable of and/or uninterested in closely charting celestial motion.

It is midwinter on Salisbury Plain in southern England 4,500 years ago. Cold and windswept, this is not a hospitable environment. It gets dark rather early at this time of year; still dozens of people have assembled there, for it is Midwinter's Night (21 December) in the megalithic monument we call Stonehenge (it means "hanging stones" or "stones on edge") (Fig. 23). Clad in warm tunics and possibly adorned with tattoos and medallions that identify their clan, men and women from several allied tribes have gathered within the 110-m (360-ft) diameter circular ditch-and-bank enclosure. They stand erect, looking not unlike the 25-ton slab shapes that make up the circle of stone that surrounds them—special worked columns they had quarried, shaped, and carted by sledge all the way from the Marlborough Downs, a distance of more than 30 km (18 miles). Semi-sedentary people come from miles around to this place during this season of inactivity, their harvest reaped, their hunting sojourns concluded. They are here to worship those who have sustained them, the gods of nature—the sun and the moon. While they gather they will also trade their goods in a market and share food in celebration of a communal festival, for all such activities go together when semi-sedentary chiefdoms assemble periodically.

The rites begin when the full moon rises along Stonehenge's main access; they know from experience when it will rise precisely in the

5-m (16-ft) high stone gateway 100 m (320 ft) northeast of the center of the circle. And they are well aware that the midwinter full moon (the one nearest the solstice) will rise opposite the setting sun, thus providing ample light for a night-long ceremony to honor the celestial deities come to earth. They may even be aware that the next full moon might be eclipsed, as the gods offer a colossal celestial spectacle to awe the assemblage.

Today only the right-hand standing stone of the 4,500-year-old gate, the Heelstone, remains,[6] but the rising midwinter full moon still keeps its ancient appointment. In the opposite season of the year, on Midsummer's Day, the June solstice, the sun itself rises over the Heelstone, marking its other extreme or standstill point on the horizon.

Modern astronomers look up. We measure celestial coordinates in a system that turns with the sky and we chart the planets by reckoning their positions along the ecliptic. Contrary to this, horizon astronomy was the dominant form of keeping track of time among cultures that have left no written record for us. One advantage to watching the astronomical horizon is clear. All it takes is a marker—a natural peak or valley, or a stone deliberately placed in a strategic location. An environmental calendar based on observations of the sun or the moon on such markers constitutes the beginning of clock making. No numbers and no notation are necessary.

The archaeological record tells us a lot about the chronology and evolution of Stonehenge. For example, we know that its earliest stages (before the stone circle was built), dated to approximately 2950 BC, were likely erected by several extended families, each consisting of no more than 50 to 100 people. They were probably in part subsistence driven—in need of a good place to hunt, to cultivate grain and later to raise cattle. There is evidence that their descendants and other new-

23 *Stonehenge. The photo shows the sun rising on the Heelstone on the morning of the June solstice, Midsummer's Day.*

comers, perhaps affected by an extended period of drought, may have mismanaged the environment to the point of crisis, denuding it of trees and offsetting the chemical balance in the soil. Still the Stonehenge sun- and moon-watching tradition would continue for over 1,500 years, lasting to a time when the culture became completely sedentary.[7]

Over the next 1,500 years Stonehenge would be constructed, deconstructed and reconstructed. First came the ditch-and-bank structure containing the 56 Aubrey Holes, likely timber settings later used as offertory pits, and the solstice gateway. The earliest structures (now long vanished) were probably made out of wood. At that time

Stonehenge may have been a simple place of assembly, with a built-in solar timer to tell people the most opportune occasion to gather together and illuminate the ritual stage. Imagine how impressive the rites to the sun god would have looked with the morning light glaring down the accessway!

By 2500 BC, the population was more settled, tending toward pastoralism and intensive farming. People lived in complexes of small rectangular houses enveloped by timber circles; they erected chambered tombs and mausolea. Some members of this society hunted on the high moors, while others mined. Scholars agree that the culture that erected the huge trilithons, which added the lunar to the solar extremes as timing devices at the horizon, must have been highly stratified, with specialized groups each assigned their own role in the project.

Of all the awe-inspiring aspects of Stonehenge, it is the astronomical tradition that captivates us. And the question we have voiced before once again comes to mind: How could these ancient people have set up precise seasonal markers without instruments and an advanced technology? Without a doubt, watching and marking the moving sun and moon with the degree of detail implied in the alignments would have constituted a full-time job. You cannot lay out all of those orientations in a frequently cloud-bound environment without a lifetime or more of careful observation and impermanent staking. We can think of these early low-tech astronomers as engineers of a sort, for they must have spent a good deal of time working at the task of laying out the alignments, correcting and recorrecting, updating and improving their precision.

Why the circular shape of Stonehenge? I think it probably derives from the conversion of a communal bounded wooden circle into a more permanent ceremonial center. The process may have resembled the enshrinement of an ancestor after death, except that in this case it was the symbol of the family, the domestic dwelling place that was elevated to architectural permanence in the ring of standing sarsen

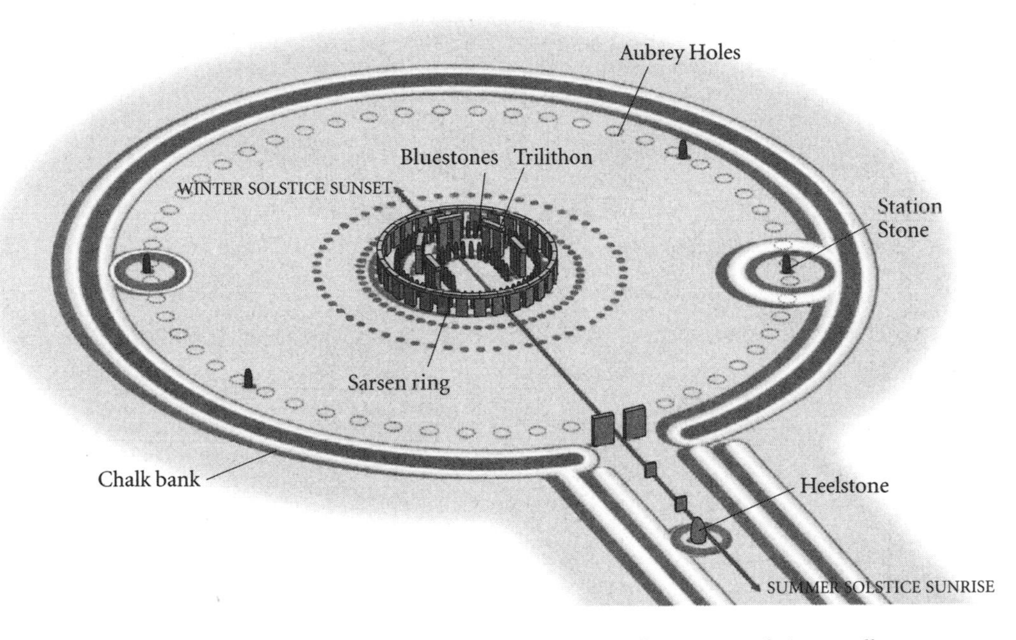

24 *Stonehenge. The main axis (lower right) points to sunrise at the summer solstice as well as full moonrise at the winter solstice.*

stones. The tradition of the woodworking capacity of these ancient people is carried on in the linteled structure of the sarsen ring and trilithons and in the vertical arrangement of standing stones which resemble the timber settings in the domestic roundhouse. We have no record of who conceived this way of monumentalizing certain aspects of their society within the single-access astronomically oriented ring structure that is Stonehenge, but it is clear that whoever they were, they had built upon the ideas of predecessors who had constructed the ditch-and-bank structure more than 500 years before. We can begin to compare these skilled laborers to the architects of the great pyramids of Egypt or of the Gothic cathedrals of medieval Europe—except that all of the latter lived in specialized state-organized societies and sought to monumentalize the glory of the gods in dazzling hand-made structures such as those we find in Istanbul, Mexico, or in Vatican City, Chartres, Cologne, Amiens, and Rouen.

Stonehenge and its sister megalithic complexes number in the hundreds and they are spread across the British Isles and northern Europe, though most are much less impressive. They were erected not just to gain access to the sky for open-air worship, but also to bring semi-nomadic people together, to conduct rites and to witness for themselves the sky gods appear in their proper places. The builders' motives were spiritual, sometimes political, always related to subsistence—all categories of human behavior required an understanding of astronomy. Think of Stonehenge, then, as a place of social gathering, a place for religious assembly, a cultic center, an economic center, perhaps a place of fortified habitations, a celestial temple, an observatory—all rolled into one; but most of all think of it as a theater designed to dazzle those who entered its inner sanctum. All these functions of the great monument cross-cut one another, perhaps some being stressed more at one time than at another. The great achievement of Stonehenge is that the genius of its inventors encapsulated a multitude of functions in a single monument. Like all people on the move, its early builders looked to the sky for their cultural template.

While such an impressive work of architecture commenced with a culture that spent a significant portion of the year migrating across a hostile landscape, the fully developed version of later Stonehenge owes its existence to an ever more sedentary culture, one of the kind to whose astronomical interests we direct our attention in the next chapter.

CHAPTER 5

The Farmer's Sky

[A farmer's] astronomical knowledge is based on rough-and-ready observations, inaccurate by their very nature, unsystematically recorded and imperfectly understood.[1]

A contemporary historian of astronomy

This sort of primitive astronomical knowledge, a farmer's or shepherd's astronomy, if you will, I should not be at all surprised to find in any settled community, whether literate or not.[2]

Another contemporary historian of astronomy

Stay in the same environment year round and watch, closely and repeatedly, what happens along the local horizon. That is all you need do to establish a precise seasonal calendar. No stylus or tablet—only a pair of sticks, or one backsight stick and a recognizable prominent point on the horizon as foresight. Use a pair of sticks (one as a foresight, the other as a backsight) to mark the place where the sun goes down; keeping your foresight stick in the same place, repeat the process next day, and the day after that. Don't worry about cloudy days. A second or third round of annual observations will fill in the blanks. Drive stakes into the ground at the places where you hold your backsight on successive dates and you will see a pattern emerge. If your baseline (the distance between back- and foresight) is long enough, you will easily note the daily difference in the sun's movement in azimuth, or distance along the horizon. (Place them a

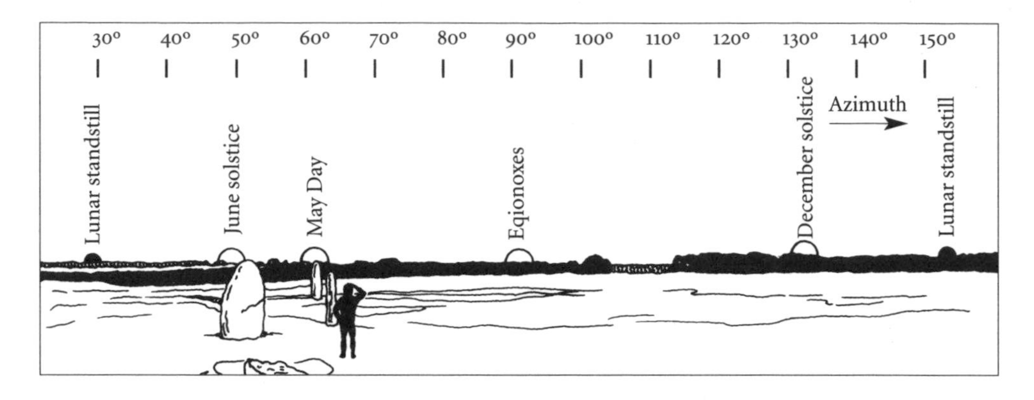

25 *Building an orientation calendar. The observer sights the May Day sunrise along a pair of standing stones that replaced earlier stakes used to lay out the alignment.*

kilometer apart and, except around the solstices, you will need to shift your backsight several meters between successive days.) You will find that sunset points drift to the right (or northward) along the western horizon from the beginning of winter to the beginning of summer; then they turn around and, after slowing to imperceptible advances around solstices, or standstill positions, retrace their path (sunrise points do the opposite) (Fig. 25).

Mark the places where bright stars rise and set. Learn with which of the sunset points the days of the first and last seasonal appearance of various bright stars coincide; memorize or mark them appropriately on the fixed sun sticks. Do all of these things and you will discover the secret of how nature's star and sun clocks mesh together. The sticks with which you mark the movements of the sun and stars are the key to the future, for repeated observation will also reveal that the arrival of the sun at certain spaces between sticks also demarcates the flood season, the dry season, the blossoming of plants, and the emergence of animals from their burrows. Set what you have witnessed in stone at a permanent location, add a bit of poetry or music, and you will have replicated the essence of time reckoning experienced in the life of the world's earliest sedentary watchers of the sky.

So why do we struggle so hard to understand our ancestors? Why do we continue to wonder how, without telescopes, their eyes could penetrate the intricacies of the visible universe? Without computers how could they precisely predict the positions of celestial bodies? Without writing how was it possible to keep records of observations on which to base accurate predictions? Without mathematics how could they calculate all those cycles? I think part of the answer is that we simply don't know enough about them; the more we dig into the record of the past the more we come to realize that our low-tech predecessors, with perspicacious eye and attentive mind, accomplished extraordinary feats.

Taken at face value, statements such as those I quoted in the epigraphs at the head of this chapter, both written by modern historians of Greek astronomy, discredit the "everyman's astronomy" on which Classical Greek and thus modern scientific astronomy are based (see Chapter 11). They assume the absence of literacy in prescientific cultures, along with their lack of precision. They deny any systematic organization to the body of data the ancient Greek farmer gleaned from the sky.

The scientific pundits could not be further from the truth, as my simple exercise with a pair of sticks demonstrates. Not only was early practical astronomy systematic, but it was also embedded in a broader world view than its contemporary counterpart—a view that encompassed cycles of meteorological and biological as well as agricultural significance.

Consider Hesiod's *Works and Days*, a 9th-century BC poem about 800 lines long, originally intended to be recited (perhaps even sung to the accompaniment of a lyre) to an audience that would have been familiar with the story: a hard-bitten farmer who works the hostile, fickle environment of the northern Peloponnese tries to educate his wayward younger brother, Perses, about the everyday knowledge of farming. To judge by what Hesiod has to say, the art of farming involved looking up as much as looking down, for every agrarian activity, properly timed and scheduled, seems to be presaged by celestial phenomena. Let's look at a few examples:

> When you notice the daughters of Atlas, the Pleiades, rising, start on your reaping, and on your sowing when they are setting. They are hidden from view for a period of forty days, both day and night, but then once again, as the year moves round, they show again, at the time you first reappear at the time for you to be sharpening your sickle.[3]

The stations of this glittering little star group in our autumn–winter constellation of Taurus (specifically the first appearance in the east and last appearance in the west after sunset) seem to have been situated conveniently so that the Pleiades functioned as a pre-emptive signal for ending the old planting season and beginning the new.

Hesiod also cites a clever little winemakers' timetable geared to the stars:

> But when the stars of Orion and Sirios have climbed up into midheaven, and rosy-fingered Dawn
> is facing Arkturos, then, Perses, pluck and bring home clusters of grapes. Set them to dry in the heat of the sun for ten days' nights, and in the shade for five days, and then on the sixth day draw off the blessings of glad Dionysos into your jars.[4]

Now modern calculations show that, around 900 BC, Arcturus confronted dawn (that is, made its heliacal rise) about 4 September, which coincided with the time Orion's Belt (followed by Sirius) was positioned high on the meridian in the south. Following Hesiod's instructions to Perses, we count ten plus six days, thus arriving at the final step in the process, scheduled for 20 September our time. One wonders whether Perses was there to help with the pouring! (Note, incidentally, an allusion to counting in units of ten in both these passages. We know that four centuries later the Greeks divided their months into ten-day segments called decades, a habit that undoubtedly originated in the pre-literate era of counting time units on the fingers.)

Some of the passages in *Works and Days* offer a glimpse into a world in which the stars are conceived as more than mere timing devices. They can affect our feelings—even arouse our passions. Says Hesiod:

> When the piercing strength of the sharp-rayed sun stops burning with sweltering heat and the rains of autumn are poured by almighty Zeus from above, and mortals would experience the change, for Sirios, the scorching dogstar, passes by day for only a short time over the heads of doom-fated men, and enjoys more of its journey at night.[5]

By my calculations the date in question is 12 September. It divides the year into periods when the brightest star in the sky is visible during more or less than half the interval of total darkness. Perhaps the early Greeks were already partitioning the days into hours. For us Sirius is still remembered as the "Dog Star" even if its appearance, thanks to the precession of the equinoxes, no longer presages the "dog days" of the long hot summer. Can any of us imagine a modern astronomer concluding that observing a bright star affects their appetite or amorous predilections?

If you want to take the trouble, as have most literate cultures, you can fashion a tabular written calendar out of the environmental cues evident in the *Works and Days*.[6] Such a calendar covers on the one hand the movement of the sun, moon, and stars in the sky above, and on the other matters pertaining to cereal grains, figs, artichokes, hay, and tree leaves, along with oxen, woodcutting, and the making of clothing here in the world below, as well as cranes, swallows, cuckoos, and rainstorms in the airy world between earth and heaven. Lived to the fullest, the agrarian life required no canonization in print of nature's ways, only repetitive, careful use of the eye that guided the hand.

There is a constant interplay between the recognition of sky events and the conduct of activities in daily life. Hesiod's astronomy is not isolated. He is not inquiring into nature "for itself." His skywatching activity makes no sense unless it is woven together with other matters of human concern. Nor is folk astronomy a subject that involves much discourse or theorizing at the scientific level, as does the later Greek astronomy of Plato and Aristotle that we'll encounter in Chapter 11. Above all, sky events emerge as expressions of the coherence and order of things. Hesiod's may not be our kind of astronomy. Yet, though devoid of instrumentation, it shares with ours the search for a systematic, ordered view of the universe around us. We are fortunate that Hesiod's orally based farmer's astronomy was later codified in print. In other instances only the mute stones of the archaeological record serve as evidence that skywatching was similarly directed in far earlier periods of civilization than previously thought.

When the Israeli archaeologist Jonathan Mizrachi asked me to examine the archaeoastronomy at Rujm el-Hiri, a 4th-millennium BC site in the politically disputed Golan Heights, the site plan he had drawn up based upon his two years of excavation immediately grabbed my attention. Rujm (Fig. 26) looked a lot like Stonehenge, though there was a significant difference in the level of subsistence of the peoples who built them.

Golan (from *Jaulan*) means "the place of wanderers or nomads," but the archaeological record confirms that by 3000 BC the plateau (situated in the biblical land near the fertile eastern shore of the Sea of Galilee at the edge of the eastern deserts of Syria and Jordan) was home to a completely sedentary community that surrounded an immense Bronze Age megalithic complex. It has been called the

26 *Aerial photo of Rujm el-Hiri, the "Stonehenge of the Levant." Like Stonehenge, its main entryway (on the northeast) aligns with mid-summer sunrise. Note the segmented compartments between the concentric rings.*

"Stonehenge of the Levant." The people who built Rujm 5,000 years ago farmed and raised goats. They lived in small villages consisting of parallel chains of houses built end to end.

The archaeological record also tells us that most of the Rujm building project was executed at one time. Nearly 40,000 metric tons of basalt went into building Rujm's nine concentric walls (the outermost is 155 m (500 ft) in diameter). Compare Stonehenge's ditch-and-bank at 110 m (360 ft), separated by seemingly randomly placed radial walls in compartments of various size. There is no doubt that precise planning went into the construction of Rujm. For example, we discovered that the builders employed a measuring unit of approximately 4.7 m (15.4 ft). When Mizrachi and I took careful measurements of the radii of the walls, we found that they all were whole multiples (2, 3, 4, 5, 7, 11, and 15) of this unit.

The situation of the northeast accessway into the site (visible at 8 o'clock in Fig. 26) reminded me of Stonehenge. Enclosed by huge boulders, the monumental aperture is almost 30 m (100 ft) wide and, like the Heelstone and Causeway of its ancient British counterpart, it aligns within approximately one sun-disk width of sunrise on the June solstice. The solstices are known to have been widely recognized among Middle Eastern cultures. They demarcated the division of the seasonal year into six-month halves or "semesters," whence our academic word. The city of Ur in Sumeria, for instance, celebrated fertility rituals at each solstice to honor the sacred marriage of Dumuzi-Tammuz to the goddess of love, Inanna. The rite assured both the fertility of the land and that of the womb of both man and beast. So it is not unlikely that Rujm el-Hiri, like Stonehenge, functioned as a ritual center and a temple in which the local urban populations worshipped.

Around 2500 BC, about a thousand years after the concentric walls and northeast aperture were erected, other builders added a burial chamber to house the remains of an individual of high status. The tomb is aligned more or less along the same axis as the accessway, perhaps a symbolic indication of the perpetuation of the fertility rite.

We discovered two colossal boulders located due east of the center of the site along the outermost stone ring; these appear to have marked the equinox sunrise, the midpoint between the June and December solstices. There is also a southeast accessway into the complex (all at *11* o'clock) but its alignment misses the winter solstice sunrise direction by a wide margin. Instead it lines up with Mt Tavor (Tabor) (likely from *tabbur* or *omphalos*, meaning navel), one of the two most prominent mountains on the horizon. A second mountain, Mt Hermon, is aligned exactly north of Rujm, so it is likely that both sky and landscape considerations guided builders in locating and orienting Rujm.

The sacred nature of the two mountains goes back at least to biblical times. For example, both are mentioned together in the 89th Psalm where a king prays for deliverance from his enemies. He praises the Lord for establishing the sun and the moon, the north and the south, and it is in this context that the two great mountains are cited: "Tabor and Hermon sing for joy at your [god's] name."[7] Mt Hermon is also called Ba'al Hermon, after the head of the pantheon of Canaan, and is mentioned elsewhere several times. Because mountains reach upward to heaven they are often considered to be the dwelling places of the gods. The gods of the Babylonians were thought to have assembled the destiny of the world atop a holy mountain, so it is not surprising that holy mounts should be configured at particular spots into the sacred landscape.

The theme of fertility in the landscape as well as in the womb has resonated repeatedly in the studies of biblical scholars who have explored mountain symbolism. Mountains are fertile because they are the major feature in biblical descriptions of the promised land. For example: "[It] is a land of hills and valleys which drinks water by the rain from heaven, a land which the Lord your God cares for";[8] "the mountains shall drip sweet wine and all the hills will flow with it";[9] "But you, O mountains of Israel, shall shoot forth your branches and yield your fruit to my people Israel."[10]

Though they were first thought to be grain storage facilities or cattle pens, there is not a shred of archaeological evidence that the compartments segmented by the radial walls between the concentric rings ever were so used. In fact, it seems possible that they always were totally empty. Perhaps the radials matter more than the spaces between them? If the radial method of sighting the June solstice

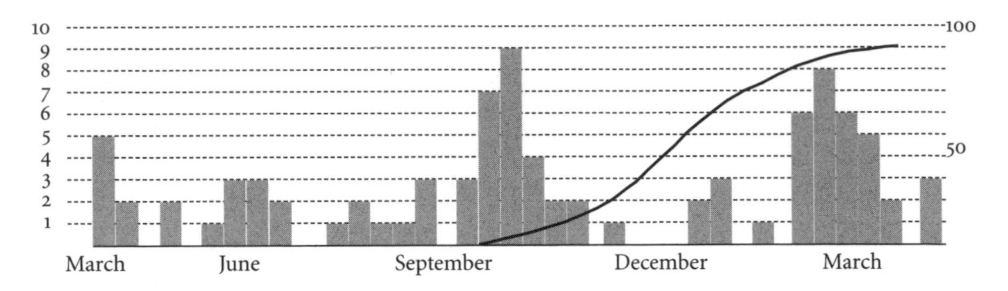

27 *Weather and alignment chart for Rujm el-Hiri. The vertical bars show the number of matches between dates of heliacal rise/set dates of stars and alignments of radial walls. The dark curve shows accumulated rainfall in cm at the right. Note that the largest number of matches fall at the very end of the dry season (the middle of the chart).*

sunrise had already been established from the center of Rujm, could the radial walls have worked in the same way? What did they point to? That the stars might have been involved is a distinct possibility. To begin with, the radials seem to avoid a wide zone centered on the northern area of the sky, where stars neither rise nor set.

While there were plenty of walls to match with an abundance of stellar targets, what emerged from our archaeoastronomical research was a preponderance of matches to stars whose appearance and disappearance dates coincided with a crucial period of climatic transition between the seasons—a time when anxious farmers would have been most likely to seek divine guidance in anticipation of the forthcoming rainy season.

Following up on a possible connection with the fertility cycle, we looked in detail at annual climate patterns in the Lower Golan. We found that the rains start in September, and while December and January are the wettest months, with rainfall tapering off sharply in March, April and May are mostly dry. The transitional period between wet and dry plays a very important part in Golan agricultural life. First the damp west wind cools the air, but this unstable period is punctuated by the sirocco (or *esh Sherqiyeh*—east wind), a hot wind that brings with it fine dust. The sirocco dries up and kills all the vegetation; it darkens the horizon and it irritates the senses. All the while, at least until the fall equinox, there is no rain.

When we matched our seasonal distribution of star alignments with a seasonal rainfall plot we discovered an almost perfect coincidence (see Fig. 27). The unstable meteorological season just preceding the onset of prolonged rainfall was clearly indicated in the sacred architecture of Rujm el-Hiri. The connection Hesiod sought to express in verse between what he saw in the sky and what happened here on earth had already been set in stone by his biblical predecessors to the east.

Little has changed in 6,000 years. Nineteenth-century Indonesian rice farmers used the stars to determine when to plant. They poured water into the open end of a bamboo pole, then pointed it at a star, thus allowing some of the water to spill out. By holding the stick vertically, they could determine when the water level corresponded to a predetermined mark. That was the time to plant (Fig. 28). They used other celestial sightings that coincided with other activities of the husband-

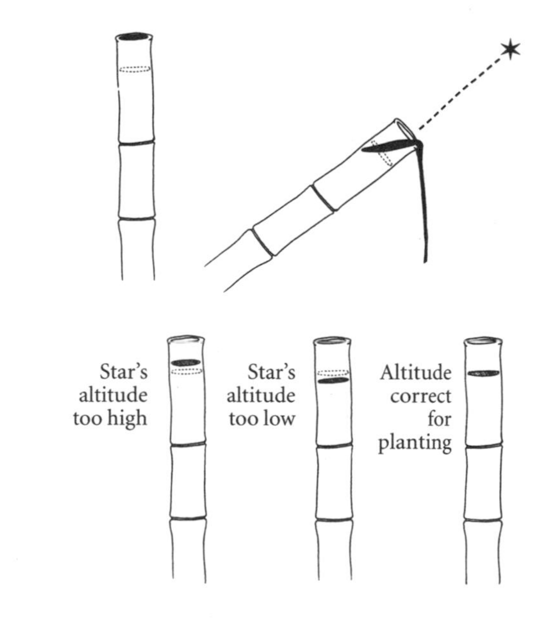

28 *The Indonesian water stick works something like a water-filled telescope. When a pre-selected star is at the right altitude the water level indicates the time to sow.*

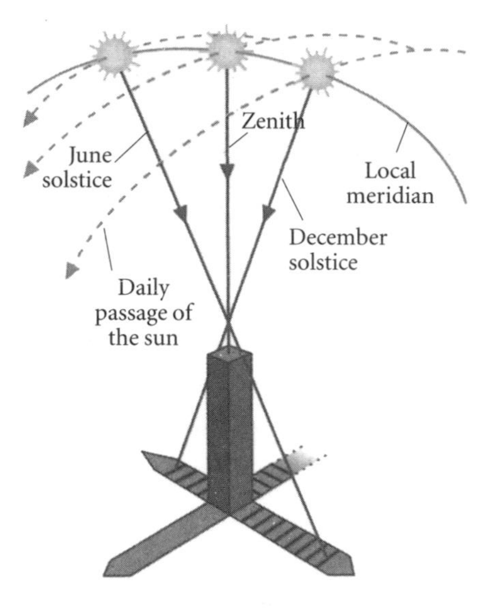

29 *The Javanese* bencet *or sundial marks out equal* spaces *(rather than equal* times*) traversed by the sun's noontime shadow.*

man, the whole of agri-time being geared to the life-cycle of the rice crop. Informants have told anthropologists that while there was once confusion about when the sowing ought to begin, in ancient times it was regulated by the stars, especially the heliacal rise of the Pleiades.[11]

The old Javanese calendar serves as an excellent example of adaptive astronomy and the domestication of time in a highly organized agrarian community. Our story begins with the discovery of a gnomon, there called a *bencet*, or *banchet*, already in use by 16th-century Dutch colonists (Fig. 29). This shadow-casting device was used to partition the year into twelve months of unequal duration, called *mangsa*, ranging from 23 to 41 days. This puzzled inquiring anthropologists until it was realized that at the latitude of Java (7° S), the *mangsa* corresponded to time intervals that would result if you used as the measurement principle equal units of length traveled by the shadow cast by the tip of the gnomon at noon. It happened because there is a very special condition that occurs in gnomonic geometry in the latitude of central Java: When the sun stands on the meridian at noon north of the zenith on the June solstice, the shadow length, measured to the south, is exactly double the length that occurs when the sun at noon lies south of the zenith at the December solstice, at which time the shadow is projected to the north. Javanese chronologists halved the shorter segment and quartered the longer, thereby producing a twelve-month calendar—their year begins at the June solstice and pivots about the two months that start with the passage of the sun across the zenith; at noon on each of these dates the sun casts no shadow. At the December solstice the shadow reverses itself.

But how does this ingenious time-parsing instrument, unique to the celestial environment in which it functioned, tie in with the approximate star-timed agrarian cycle? The descriptions of the duties of the husbandman acquired by early Dutch anthropologists read like a page out of the "Works" section of Hesiod's *Works and Days.* As in Hesiod's description of husbandry in Classical Greece, we discover that the organized work activities of the Javanese farmer complement natural events. For example, the farmer works especially hard at mid-year to facilitate the growth of his staple crop before all can be left to mature in the later segment of the seasonal cycle. While the commencement of the various activities was once casually determined, the schedule later came to be designated by a more formal calendar. The starting times were determined by the priests of various villages, each of whom consulted the local *bencet.* If you calculate gnomonic shadow lengths in the range of latitudes of Java, as I have done,[12] you will easily discover the symmetry native chronologists must have arrived at to set up the standard *bencet* that was used to regulate the agricultural calendar. Like the Greenwich meridian, the *bencet* created an island-wide "standard time" that local farmers could use to adjust their natural clocks. But nature's symmetry is strange, for the month lengths, while equal in space, turn out to be decidedly unequal in time.

Although this exotic-looking calendar offered the distinct advantage that it placed the longest months of the year in the season demanded by the needs of labor, it was still not accurate enough to fit local needs, which varied across the variable climate zones on the large islands. Like the Mursi (whom we met in the last chapter) and the Trobrianders, local calendar keepers felt free to slightly distort the rhythm of time ordained by the standard *bencet*; nonetheless their representatives always consulted it. (Our own calendar months, originally based on the lunar phase cycle, were similarly distorted in the time of the Roman empire to meet the more practical needs of the seasonal calendar—see Chapter 10.)

Other changes in the calendar followed Indian influence, which came to the island in about the 9th century AD. When the lunar year of the Hegira was introduced, the seasons gradually slipped out of joint with the *bencet*-timing mechanism because of the eleven-day lapse between the lunar and solar years discussed in the previous chapter. The first ten months carried the names of the ordinal numerals in the local language; the last two were probably added later by the Brahmins to force the calendar to correspond to their own. Still later the Brahmins threw back the beginning of the year from the first to the eleventh month, which corresponds roughly with our April. They did that in order to make the native time cycle fit the Hindu year. A mystery still remains, however, for we are not aware of the exact origin of the idea of a solar zenith pivot and a gnomonic device that tied the regional calendars together.

The farmer's calendar persists in modern folk wisdom, far from the fringes of contemporary scientific agronomy. Some organic farmers today still look to the sky to align their actions with nature's cosmically timed cycles. According to one theory gardens grow best if planted in harmony with the phases of the moon. For example, plant broccoli, lettuce, annual flowers, and herbs (those that produce their seeds on the outside) during the first quarter of the lunar cycle. Vegetables that set seeds within a pod or skin—that is, beans, tomatoes, and squash—have an affinity for a second quarter moon, so plant them during that phase of the lunar cycle. Vegetable root crops, such as potatoes and onions, grow best if placed in the ground during the third quarter of the cycle, while garden cleanup prefers the end of the phase cycle, the fourth quarter; believers say weeds pulled then will not grow back.

Their logic goes like this: both the sun, which regulates the seasonal cycle, and the moon, whose phases are timed in accordance with its position relative to the sun, are responsible for the oceanic tides. When the sun and moon line up with the earth, at new and full moon, the tides are greatest (that is, high tides are highest and low tides are lowest), and when they lie at right angles to one another as seen from earth (first and third quarter), the range of the tides is minimal. Not even the sternest skeptic will dispute the reasoning to this point. But where claims of efficacy are based on minuscule variable effects the paths of modern scientific rationality and old folk wisdom part ways. (Homeopathy in medical treatment is an example. It is based on the principle that the tinier the dose of a particular substance, the greater its curative effect.)

The logic continues: just as the celestial bodies produce the tides, so, too, they allegedly affect water in its more subtle condition, pulling variably on water in plants, causing moisture to rise and fall in the soils. Seeds will absorb more water at full (and new) moon. Moonlight also enters the equation. Just after full moon the gravitational effect of the tides is still considerable, but the intensity of lunar light begins to wane. Energy is drawn to the roots, which is why you should plant your root crops between the full and third quarters.

The same holds for fishing. Solunar (sun-moon) tables list best times for hooking fresh- and saltwater fish. Since the feeding habits of most fish (at least those of the sea and estuaries) change with the tides, this makes sense. Turning tides affect salinity, which in turn influences feeding habits. One experiment showed that oysters transported from an Atlantic harbor to a lab near Chicago adjusted the times of opening and closing of their shells to tides that would have occurred in an ocean at the latter location, if one existed. But tides along the coast can happen earlier or later than average depending on site location and water depth.

Morning and evening twilight have long been known as good times for fishing. Moonlight matters here, too. Certain species of coral are known to breed by the light of the moon, laying down monthly growth ridges in accordance with the lunar cycle. And some full moons rise higher in the sky during different seasons of the year. All of these functions and factors—and some claim many more—conspire in the creation of complex solunar tables for specific geographic zones and times. Many fishermen, some of whom employ the temporal tabulations down to the nearest ten minutes, swear by them. Others are not so sure. An inquiring fisherman who logged his catches over a long period concluded that it all boiled down to simply noting the four 90-minute windows surrounding the daily rising and setting times of the sun and the moon.

Skeptical gardeners outnumber skeptical fishermen. They cast an especially wary eye on believers in astral energies that impinge on all life forms, thereby dramatically affecting their behavior. All biorhythmic theories are based on numerology, anecdotal evidence, and media hype, say the skeptics.[13] They contend that not a single scientific study, free of methodological and statistical error, supports any of the claims.

We might also ask: Did ancient Greek farmers like Hesiod and their biblical predecessors in the Levant really believe that the stars caused changes in weather patterns? Aristotle (384–322 BC) once claimed such a connection. But later philosophers (for example, Geminus in the 1st century BC) pointed out that the stars only served as indicators of meteorological conditions which the farmers had previously experienced: the wind from the north in summertime is announced by the morning rise of the Dog Star, Sirius, and the evening appearance of the star that represents the tip of the wheat staff in Virgo (γ Virginis) signifies the beginning of the rainy season.

Having written a book on such collected folk wisdom[14] I can testify that there never will be a meeting ground between it and science, for while one operates in the realm of the logic of natural law and the statistics of coincidence, the other is based on an abiding faith in one's personal experience. There is no scientific retort to "It works for me." Laying aside the never-ending battle between modern science and the paranormal, the fact remains that vestiges of fertility rites as old as those hinted at by Hesiod and likely held at Rujm el-Hiri live on in our contemporary seasonal calendar.

For example, May Day once opened the second half of the Celtic year, dedicated to the explosion of activity that unfolds when the life forces of summer arrive to claim victory over the previous winter's death and gloom. Possibly connected with the Roman festival Floralia, which honored the goddess of flowers ripening into fruits (and consequently the pleasures of youth), the day was celebrated in medieval England by dancing round a Maypole. (A stodgy 17th-century evangelical Protestant cleric tried to outlaw the practice, declaring the pole a phallic symbol, a remnant of Roman worship of Priapus, god of male potency.)

Long neglected because of its 20th-century association with Marxism, the essence of the old May Day holiday today enjoys something of a revival. Schoolchildren celebrate a festival of joy and the coming of summer with activities to spice up the curriculum—like reading Robin Hood stories, adorning trees with scraps of yarn and colored paper, decorating shopping bags with May Day flowers and pinning them to the board, and taking home notes asking parents to contribute coins to be placed in the May Day wishing wells for donations to the poor.

In addition to May Day, there are three other quarter-year time posts, located midway between solstices and equinoxes: Ground Hog Day and Candlemas (1–2 February), Halloween, All Saints' Day and All Souls' Day (31 October, 1–2 November), and Lammas (1 August).[15]

Lammas used to be a red-letter day (days so named either after the color of the cloaks of the bishops who attended to the calendar, or the color of the ink they originally used to pen in the special days), but its color has faded as society strays further from its old ways. The word means "loaf mass" in old English. It began as the first in a series of 9th-century English festivals celebrating the harvest, when breads made from the season's first crop were blessed, broken, and offered to the four corners of the domicile for protection.

Modern farmers' markets, devoted to sharing the new abundance of the land, are a vague remnant of the custom of displaying the first of the crops. Still, the spirit of Lammas lives on in those of us who always manage to be present for nature's "firsts," whether we attend a strawberry festival in the U.S., the new corn ceremonies of Native Americans, an asparagus festival in Germany, or the celebration of the arrival of the new vintage in wine-making regions; America's Thanksgiving is also a survival. When we dine on new potatoes, native corn, or the first salmon or oysters of the season, we all collectively participate in the rite of bringing back the first of its kind in agrarian life's seasonal round. Practically all our seasonal holidays derive from key points in the cycle of subsistence farming and they were once timed by cycles in the sky—a precious legacy washed over by techno-time.

CHAPTER 6

The House, the Family, and the Sky

The hogan is built in the manner of this harmony. The roof is the likeness of the sky. The walls are in the likeness of the Navajo's surrounding: the upward position of the mountains, hills, and trees. And the floor is ever in touch with the earth mother.[1]

A Navajo native

The Skidi were organized by the stars; these powers made them into families and villages and taught them how to live and how to perform their ceremonies.[2]

An early 20th-century anthropologist

If yours is a typical suburban home it is probably located on a sinuous street not far from a major highway. Should you live in a slightly more urban—and therefore likely more vertical—environment home base may find its place in a randomly skewed grid plan that requires a third coordinate to specify its precise locale. Your house—again if it is typical—will have a spacious dining room reserved for special occasions, even if today's version of "dining" often takes place on an individual basis, and at any hour, and in any room of the house. Your residence also will have a "living" room (also inappropriately named). It used to be called a parlor (from the French word *parler*, meaning "to speak," literally a room for conversing). It is possible you could have more bedrooms than you really need, a private den or office for each breadwinner, and a game or recreation room for pure entertainment. If your house comes equipped with a sun porch, it is probably positioned at the back of the house, regardless of where the sun travels in the local environment. The porches of a century ago used to be attached to the front of the house, where residents could exchange greetings with strolling passersby.

To judge from its architectural plan, today's domicile is less communally arranged than it used to be. Each inhabitant, though a member of the family, often has a private space within the sanctuary. Environmental factors play the scantiest role in the placement and orientation of the house. Water can be piped in from afar, trees of all kinds can be planted, and you can pay extra for a mountain view. Where are the sun, moon, and stars in relation to your living space? Who cares?

But there were—and still are—societies for which heaven mattered, often providing a template when it came to setting up the structure of the household. Take the cosmically based plan of the Navajo dwelling hinted at in the first epigraph to this chapter. The Navajo hogan is more than a home. (The word derives from *ho(o)* = "place" and *ghan* = "home.") Navajo elders say that the first home place, the hogan of creation, was constructed at the edge of the world, near where the creator gods emerged into the world of the present (recall the sky story of Navajo creation, the *Diné Bahané* narrated in Chapter 1). It was in that hogan that Black God made the stars and put them in the sky. And so, out of reverence to him, all hogans must be oriented in a direction sympathetic with the stars' movement across the night sky. The roof of the hogan must be peaked or domed like the sky, and it must be round like the shape of the sun (*ha'a'aah* meaning "the round object that moves in regular fashion"), the source of light and heat. And above all it must face east (Fig. 30).

Every hogan has four posts, one positioned in each of the cardinal directions, each mimicking one of the four mountains that hold up

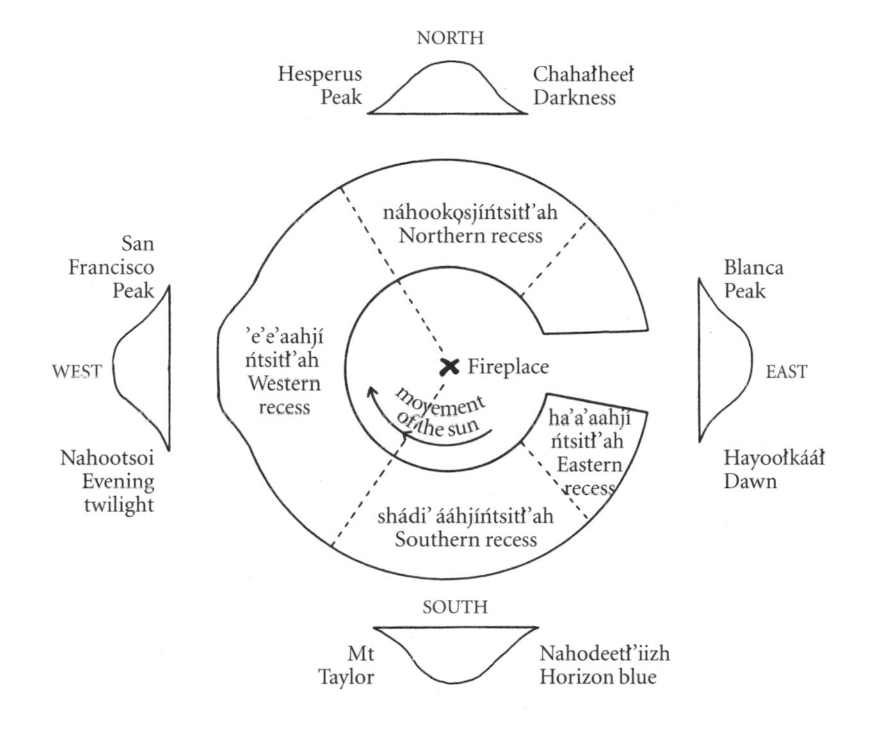

30 *The Navajo hogan is patterned after the universe around it, as shown in this schematic.*

the sky. The walls of the hogan are vertical, just like the mountains. When you enter you must move clockwise imitating the sun's motion. The interior is not physically divided, though four specialized areas or recesses—north, east, south, and west—are recognized, along with a fifth direction, the center, which symbolizes the sky that surrounds the central fireplace, the sun.

Suppose we make a few analogies between the interior space of the hogan and the modern home I described earlier. The center would be the kitchen and dining room, the place where you cook and eat; the west would be where you entertain visitors in the daytime, where elders tell stories in the evening, and where you sleep at night. The south side would become the place where you work, at weaving or making jewelry—things you do during the daytime. Most of us do not give over a major portion of our living space to prayer, but in the Navajo hogan the whole of the north recess—fully one quarter of the hogan's interior—is intended for reverence. It is a place to contemplate, to prepare for ceremonies. In the words of one elder (my first epigraph):

> The hogan is … the shelter of the people of the earth, a protection, a home, and a refuge. Because of the harmony in which the hogan is built, the family can be together to endure hardships and grow as a part of the harmony between the sacred mountains, under the care of "Mother Earth" and "Father Sky."[3]

Isn't this what a home *ought* to be?

The Skidi Pawnee built their lodges on the plains of Nebraska. They built them big enough and high enough (typically 15 m (45 ft) in diameter and 5 m (15 ft) high at the center) so that those who lived there could also feast and celebrate rites, even teach their young in them. Like the Navajo hogan, the Skidi Pawnee lodge mimics heaven's plan. The domed shape of the lodge (Fig. 31) imitates the perceived shape of the sky, down to the precise arrangement and orientation of its components, with the doorway on the east, four directional support poles, and a carefully positioned smoke hole through which to view the stars. As you can see by comparing Figs 30 and 31, the cosmically based structure of the hogan is not so different. An early visitor who inquired about the lodge's structure was told that the supporting posts are really the two Morning Stars in the East, and the two inner poles are the messengers to the Morning Stars and the Black Meteor.

The Pawnee lodge is also a functioning calendar. The doorway is

31 *Inside the Pawnee lodge: The dome shape represents the structure of the heavens, and its circular plan imitates the horizon. Posts supporting the dome symbolize sky objects. The doorway is positioned so that the sun, when rising at the equinox, illuminates the altar at the rear of the lodge.*

precisely positioned so that during two periods of the year—the twenty days that span the equinoxes—it admits sunlight in such a way that it reaches an altar on the extreme west side of the interior. This bright shaft of noontime sunlight that enters the smoke hole has another use in that it changes with the seasons. It comes part way down the wall at winter solstice and reaches all the way down to the floor by mid-February. The points where the sun touches the boundary between the wall and the floor can be mentally scaled out to

denote different times of the year—a true wall calendar! By the time the inhabitants abandon the winter lodge and move to the outdoor *tipi*, the solar image will have migrated to its position closest to the center of the lodge.

Furthermore, star groups recognizable from Pawnee lore can be sighted in the house's apertures. For example, the Pleiades,[4] a Pawnee symbol of unity, are first glimpsed briefly just before sunrise in late July through the smoke hole by an observer sitting at the wall along the lodge's axis of symmetry. They are seen again just after sunset around the time of the winter solstice. The Chiefs in Council (our constellation of Corona Borealis) shine through the smoke hole in direct opposition to the Pleiades' appearance. This may explain their opposed location in space on native star maps.[5]

Maybe the Pawnee lodge seems more like an astronomical observatory or a planetarium than a house to you. In fact, these shelters from the cold winter air once were nature's living classrooms. There one could experience star and sun scenes that dramatize the real-life "just-so" stories (like Rudyard Kipling's fantastic, easily remembered accounts of how natural phenomena came about) and moral tales told within its warm confines. No need for powerpoint here! But how do we know 19th-century occupants really made observations through the skyhouse apertures? In his book on life among the Pawnee, written at the turn of the 19th century, the early anthropologist James Murie wrote: "At a certain hour, a priest ... looked up through the smoke hole. If he could see the seven stars directly above, it was time for the planting ceremonies."[6] Unfortunately Murie gave few details.

The idea of constructing a heavenly house goes far beyond the confines of native North America. For example, the Warao, who live in the Orinoco River delta at 9°N latitude, think of their world as a flat disk that floats on a salty sea. A bell-shaped celestial canopy is connected to the disk of the world by the axis mundi, a serpent of creation which in their divinatory rites is represented by hallucinogenic tobacco smoke coils. The Warao give special emphasis to the zenith, which is no surprise (recalling Chapter 3) as they live very close to the equator. Four petrified trees that mark the cardinal directions hold up the disk. The round houses these people still build today correspond very closely to the details of this cosmology.

Reminiscent of the Pawnee lodge, the communal living quarters of the neighboring Yekuana (Fig. 32) feature a single roof window eccentrically placed to receive sunlight during the dry season and to provide an effective outlet for smoke. The window (shown in the open position in the photo) also serves the practical function of marking out the yearly calendar as its shadow cast by the sun ascends along the central pole of the house. At the same time it fulfills the religious duty of following the sun god to his celestial home at the very

32 *The canopy of heaven served as an architectural model of the wattle-and-daub roundhouse in the Yekuana culture of tropical South America. Like the Pawnee lodge, it features a roof window (shown open on the right) that receives the sun. The path of the sunlight on the inner wall serves as a seasonal calendar.*

top of the smoke-filled world axis, where the creator gods play a game that makes the seasons go by.

The Desana people who live in the rain forests of the Amazon and Orinoco (today Colombia and Venezuela) build their longhouses according to a six-sided plan. Informants told one anthropologist that each vertex of the hexagon consists of a housepost that can be identified with one of the basic support stars. The bisector of the hexagon on earth is a ridge-pole identified with the Pleiades, which today rise in this area just after spring equinox, thus signaling the beginning of the main fruiting season. To commemorate the event piles of palm fruits are heaped at the center of the house, where Orion, symbol of the center, dwells, no doubt because he lies in the middle of the astro-hexagon and because his belt perfectly straddles the equator. Once again he is a hunter, but unlike his Old World counterpart, Orion of the jungle is far more ambivalent. He possesses many incongruent character traits: ancestor and hero on the positive side, sinner and victim on the negative. (Recall the Carib version of his story recounted in Chapter 2.)

The Desana year is divided into two rainy and two dry seasons. They reckon the central point of their calendar by the place and time where the shaman's staff will cast no shadow when held upright. Like a horizontal lid, a celestial hexagonal template consisting of the bright stars Procyon, Pollux, Capella, Canopus, Achernar, and one of the stars of our constellation Eridanus (τ) overlies the earth at sunrise and sunset, just as the sun is positioned at the equinoxes. At this time, when heavenly symmetry is in force, a vertical shaft of sunlight is said to fall on a mirror-like lake below, thus fertilizing the earth. Furthermore, the original tribes were said to be six in number and they still organize themselves socially in a hexagonal model.

Just as Shakespeare contrasted the stages of life with the progress of the seasons, the Desana mark out youth, maturity, and old age on their sky hexagon, which they bind to their terrestrial six-sided home.

Expressed in a dance, a symbolic journey around the longhouse typifies the cyclic journey of both men and women through life. Each of the stellar vertices represents a significant marking post along life's road. Men, for example, move clockwise from Capella (naming) to Pollux (initiation) to Sirius (marriage). Women travel counterclockwise, but only until they arrive at Sirius; then they turn about and join their husbands. When all return to their starting point, the Pleaides and Aldebaran, they are reborn—precisely on the equinox line.

A continent away in Africa, the Batammaliba of Togo and Benin, inheritors of one of the great art-producing dynasties of pre-contact Africa, are among the master cosmic builders whose architecture has been studied in some detail. They structure their world after the universe erected by their gods. Their architecture exhibits a direction toward the sun—in holy temples and tombs, as well as in the structure of the domicile (Plate 3).

Because Kuiye (the sun) was and is always there, the Batammaliba reason that no power could have ever created this deity. To make their own house just as everlasting they carefully align its cross-beams to point to the direction of the equinox sunrise and sunset, the butt end to the east, the narrower end to the west. This enables Kuiye's rays to fall on the shrines of the deceased elders so that he may speak with them regarding the affairs of the living who occupy the house. The Batammaliba say that Kuiye is the one to talk to, for he created us and he protects us as well. We never see Kuiye—only the reflecting mirror positioned on his forehead. The real sun is human in form and he lies beyond the sky. His house is in the western sky but the doorway points eastward. This is why all shrines must open to the west to face his residence—so that we can commune with him. Like the calabash, a round or oval gourd which the Batammaliba relate to the shape of the world, their houses are circular and have a raised terrace topped by a circular roof. So, in effect their house is Kuiye's house, the house of the ancestors. They are privileged to live in his space, they say.

As in the Pawnee lodge and the Yekuana roundhouse, cosmology intersects with astronomy in the Batammaliba home. "House-horns" located over the door (they are visible in Plate 5) refer to the number of days it took the creator to make the world, and the conical mounds in front of the door commemorate the piles of stone he used to build up its structure. Newly built structures are consecrated by inverting half a calabash in front of the doorway, a symbolic template to trace out the flat horizon line of creation on the ground. A similar tracing is enacted around the entire Batammaliba village in a funerary ritual; the village "Earth priest" uses the feet of a sacrificed goat to circumscribe a line about the basically circular village, which consists of several dozen houses.

For the Batammaliba, like all the other people cited in this chapter, designing a home involves much more than simply managing solar light and heat. Every time a house is finished the builders honor Kuiye. At noon, when the sun is in the center of the sky (highest on the meridian) they place cooked cereal on the hearthstone in the center of the home.

While the North American Pawnee and Navajo, the South American Yekuana, and the African Batammaliba say the sky is in your house, remote island dwellers in the mid-Pacific say your house is in the sky. The people of the Gilbert Islands (latitude 3°S) segmented the heavens into named zones, formed by slicing the celestial sphere several times in the vertical direction (parallel to the east–west line) and by cutting it with another set of lines running parallel to the horizon (*te tatanga*) (Fig. 33). (Recall the setting of the tropical sky discussed in Chapter 3, especially Fig. 16.) The vertical lines they called the ridge poles (*te taubuki*, corresponding to our great circle of the meridian) and the rafters, segments of smaller circles (*oka*), while the horizontal arcs of the celestial sphere signified crossbeams or purlins (*te marena*) that supported the rafters of the great cosmic house in which we all live.

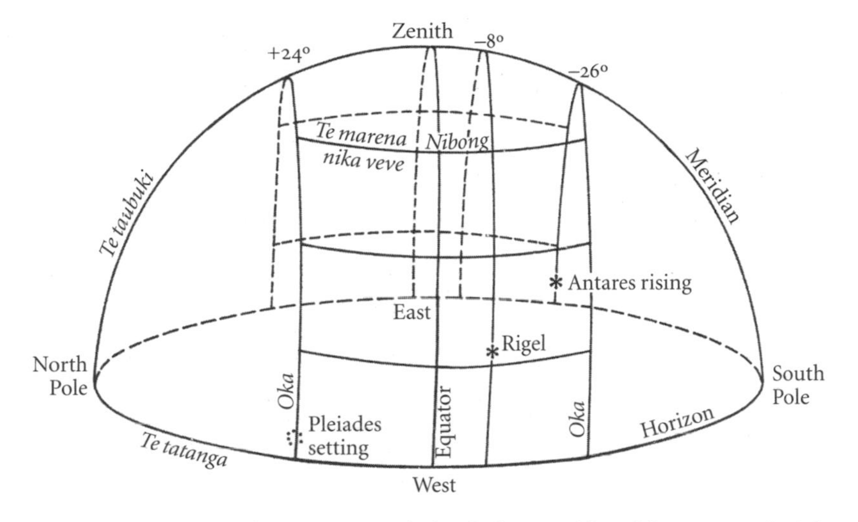

33 *The Gilbertese sky dome is patterned after the house, with each important celestial coordinate named after the corresponding part of the architecture of the house.*

In this sky-house analogy, locations of stars were described in terms of their position in one of a collection of imaginary boxes that divide up the celestial sphere, and these in turn render it a natural clock. For example, when the Pleiades reached the first purlin in the east (at an altitude of 22½°) an hour before sunrise, they knew that the sun lay at the June solstice. A similar sighting of Antares signaled the arrival of the sun at the vernal equinox. As one informant told a 19th-century anthropologist: "When you see *Rimwimata* [Antares] in the middle, between the ridge pole and the first purlin to westward, you know that the sun is on his *bike ni kaitara* [island of face-to-face], their term for the vernal equinox."[7]

When expanded to a larger domain the sky-house becomes the sky-village, as my second epigraph suggests. Before they were removed to reservations in the state of Oklahoma in the 1870s, the Skidi Pawnee lived along the Platte River in what is now Nebraska. Their villages were dismantled, but informants who once lived in them told early 20th-century anthropologists their divine plan, manifested by a deeply rooted celestial cult that believed all power resided in the stars.

Once, whole Skidi villages were arranged relative to one another according to the positions of the stars, each village having been assigned its own patron star, and in turn being charged with the primary responsibility for conducting religious rites associated with it. Thus ceremonies at the shrine of the village farthest to the west, under the rule of Tirawa, god of life and knowledge, were the first in the calendar to be performed. The initial rite, called the First Thunder Ceremony, happened when springtime thunder was first heard around the time of the equinox.

The four bowl stars of the Big Dipper, pivoted about the North Star, mapped out the positions of the five central villages on the ground. Their shrines represented the affairs of the people. Each village supervised a ritual connected with tribal matters, such as planting, harvesting, hunting, installing leaders, and conferring honors on warriors. As the order of the seasons progressed, the rites shifted from shrine to shrine, beginning with the westernmost village hosting the First Thunder Ceremony. The eastern village, under the morning star, came last in the sequence. Its function was to conduct a sacrifice that connected the world above with the world below, and to insure the perpetuity and productivity of all life on earth. Then the cycle would repeat itself (Fig. 34).

We don't know much about what went on during these ceremonies. Pawnee leaders were fairly secretive about how they were conducted, but one informant detailed that they "gave an account of creation, the establishment of the family, and the imagination of rites by which man would be reminded of his dependence on Tirawa of whom he must ask food."[8]

The sacred rites that followed the celestially ordained village plan also reminded Pawnee people about the duality of the universe. The stars in the east were male while those in the west were female. Like the Iroquois, who also fashioned a gendered sky, the Pawnee thought that the central star Polaris divided the male and female segments of the sky along a line connecting the Corona Borealis on one side and the Pleiades on the other.[9] The Pleiades ruled the midwinter sky, when women's household activities dominate and when Mother Earth began to reawaken, while Corona Borealis dominated the midsummer sky, when the political activities of the male world were prominent. Pawnee leadership rotated annually among the four central shrines, beginning with the yellow, feminine star of the northwest, then passing to the red, masculine star of the southeast, the white, feminine star of the southwest, and the black, masculine star of the northeast concluded the four-year cycle.

Given what we know about the cosmic foundation of the Pawnee village, the earth lodge can be thought of as a microcosm of the village in the sky. Its four posts represent the four central villages. The star of the east is the doorway that opens in that direction and the little shrine at the back of the lodge represents the star of the west (again see Fig. 30 on p. 113). Had English settlers never come to the New World, had the Pawnee not been uprooted, had they consolidated their influence over surrounding tribes, indeed had they built a great city to reinforce their power and influence, it is quite possible that the

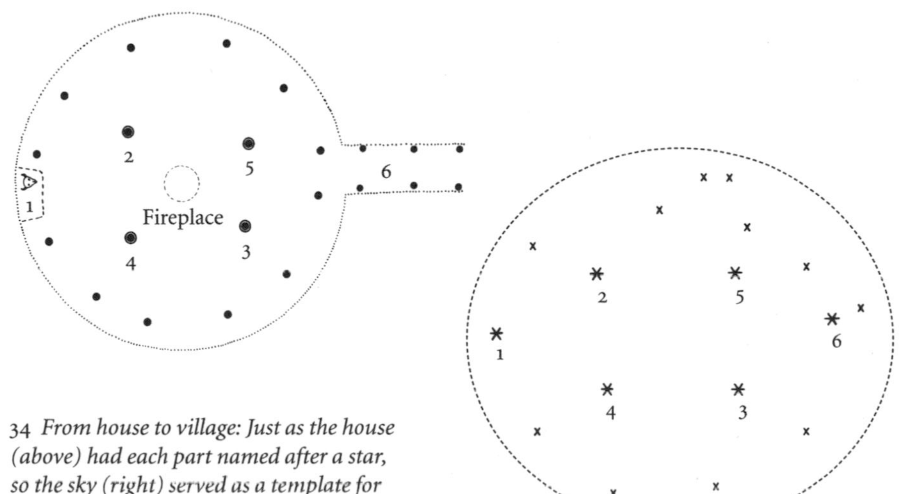

34 *From house to village: Just as the house (above) had each part named after a star, so the sky (right) served as a template for the entire Pawnee village.*

plan of that city would have retained and perhaps even elaborated upon many of the cosmic hallmarks that governed the sacred space of the house and the village. This never happened on the plains of Nebraska, but it did occur in a fertile valley at an altitude of 350 m (1,150 ft) in the mountains of Peru over 500 years ago.

The Incas dominated the Andes and the coast of South America from Ecuador to Argentina during the 14th and 15th centuries, prior to Spanish contact. As we shall learn, the sky was very important in their culture. Of all the reasons we can offer for the success of the Inca empire, the most persuasive are the strict order and the high degree of organization built into every component of it. Early Spanish visitors to the highland capital said that nowhere in Peru was there a city with such an air of nobility as Cuzco; by comparison the other provinces of South America were merely settled, lacking design, order, and wise rulers to command them.

The Incas called their empire "Tahuantinsuyu" or "the Four Quarters of the Universe," because of its basic four-part plan. They segmented it into *moieties* (complementary social subdivisions) called Hanan (higher-ranking, or up-river) and Hurin (lower-ranking, or down-river) Cuzco. Each half in turn was split into two sectors, or *suyu*. The lines that separated the *suyu* demarcated the flow of underground water in the Cuzco valley, which naturally follows a radial plan in this montane environment. The *suyu* served as an organizing principle that defined water rights for the various family groups who farmed the wedge-shaped plots of land between the river valleys. The Incas believed they received the underground water by right of birth directly from their ancestors, who resided in the body of the earth mother (Pachamama), and whom they honored and nurtured by making sacrifices to feed her.

But the relationship between kinship and geography did not stop there. The so-called *ceque* system unified Inca ideas about religion, social organization, calendar, astronomy, and hydrology. A 16th-century Spanish chronicler has left us a thorough and detailed description of this unique topographic mnemonic. It consisted of some 41 imaginary radial lines (*ceque* = ray) that cut across the landscape and were grouped like spokes on a wheel according to their location within each of the four *suyu*. The wheel's hub was the Coricancha (the Temple of the Ancestors).

Family lies at the center of the system, for the chronicler tells us that each *ceque* was assigned one of a set of three hierarchical groups that represented the social classes that tended to it, and maintained the 328 shrines (*huaca*) that demarcated the *ceque* like so many beads on strings. These were (a) those *ceque* said to be maintained and worshipped by the primary kin of the Inca ruler; (b) those that were worshipped by his subsidiary kin; and (c) those tended to by that segment of the common population not related to the ruler.

The idea that the forces of nature are divided among kinship classes is actually fairly widespread. Recall that in ancient Sumeria each tribe and town had a special sky deity who watched over them. And in parts of aboriginal Australia the night sky was regarded as a pair of great camps divided by the river of the Milky Way. Stars on one side were named for kin classes in the Aranda camp, those on the other side for such groups in the Luritja camp. Marriage alliances between the two were reflected in the star-naming process. Thus α and β Crucis of the Southern Cross were said to be the parents of nearby α Centauri, α Crucis being the male of one kin class, β Crucis the female of another. They live west of the great river, while α and β Trianguli, along with their son β Centauri, live in the east—just as their counterparts do here on earth.[10]

In the Inca *ceque* system, the assignments on the hierarchy of worship of the 41 *ceque* lines rotated sequentially (a), (b), (c); (a), (b), (c); etc., as one proceeded from one *ceque* to the next within each *suyu*, passing all the way around the horizon in a clockwise direction in the northern *suyu* and counterclockwise in the south.

The shrines of the Inca, *huaca,* could be natural rock or man-made temples, intricately carved rock formations, bends in rivers, fields, springs or other natural wells (called *puquio),* hills, even impermanent objects such as trees. At each of the *huaca* the designated worshippers communed with the gods who controlled the cosmic forces. There they left their offerings. To give an impression of the detail of complexity in the system, let me cite an example (among the 328 described by 16th-century chronicler, Bernabé Cobo), which also introduces a celestial element into the shrine:

> The seventh [*huaca* of the eighth *ceque* of the Chinchaysuyu, or northwest, quadrant of the city] was called Sucanca. It was a hill by way of which the water channel from Chinchero comes. On it, there were two towers as an indication that when the sun arrived there, they had to begin to plant the maize. The sacrifice which was made there was directed to the sun, asking him to arrive there on that hill at the time which would be appropriate to planting, and they sacrificed to him sheep, clothing, and miniature lambs of gold and silver.[11]

The rich detail offered in this passage suggests that the *ceque* plan must have been an important part of life in the old prehispanic capital. There is some rather concrete information about the flow of water and its connection to the celestial events present in it as well. Other passages connect observations of the sun made from each of the *moieties* to respective sets of pillars erected on their local horizons, all these observations being tied to the regulation of a unified orientation calendar.

The system of observations was arranged so that each segment of the community could play a role in the domestication of time. To judge by descriptions given by the Spanish chroniclers, the pivot of the calendar may have been the date that marked the point in the year *opposite* the solar zenith passage. Thus, an anonymous chronicler describes the passage of the sun by four pillars on a hill called Cerro

Picchu, which overlooks Cuzco from the northwest. His description fits with that of Sucanca in the previous passage, even though Cobo mentions only two towers there:

> When the sun passed the first pillar they prepared themselves for planting in the higher altitudes, as ripening takes longer.
>
> When the sun entered the space between the two pillars in the middle it became the general time to plant in Cuzco; this was always the month of August.
>
> And when the sun stood fitting in the middle between the two pillars, they had another pillar in the middle of the plaza, a pillar of well worked stone about one estado high, called the *Ushnu*, from which they viewed it. This was the general time to plant in the valleys of Cuzco and surrounding it.[12]

Our investigations suggest that the observer in charge of the calendar likely stood somewhere in the present-day main plaza of Cuzco, the Plaza de Armas, a few hundred meters from the Coricancha, to watch the sun set. The time was mid-August. The Inca discovery that the important agricultural date connected with the coming of the rains and the planting season happened to coincide conveniently with a significant solar phenomenon in the environment of Cuzco, may have led them to see the act of planting and the passing of the sun underneath the world as like-in-kind events. At this time, an Inca legend says, the earth mother "opens up." Pachamama is then at her most fertile and can be penetrated both by the tiller (with his plow) and the sun (with his rays). The idea of a line of sight that connects the rising and setting sun on the days of overhead and underfoot passage may have been a way of expressing the vertical nature of the ecology of the Andes, but in horizontal space. In other words, as the sun advances horizontally across the towers, the time to plant advances vertically.

Like the design of the Navajo hogan and the arrangement of the Pawnee villages, Cuzco was an entire city designed to replicate the order in the universe and to delineate the pathway of time. As in the other examples discussed in this chapter, it was also a means of structuring social order. The whole system worked when each particular kinship class performed its assigned function in the proper place along its *ceque* line and at the correct time designated in the calendar. In this remarkable meeting of nature and culture in a harsh and variable agricultural environment, the *ceque* system was a highly orchestrated scheme devised by Inca royalty to prescribe proper human action—a plan based upon residence and kinship in a radial, four-fold geographic framework. As in the other examples discussed in this chapter, sky events were a major component of that order, but they were inseparable from all the other components. And so we dare not try to pull it out of context, for then it will make no sense. As we shall discover in the next chapter, Cuzco is but one example of what can happen when a single village with a divine plan based on celestial movement develops into a highly influential city or empire.

CHAPTER 7

The City and the Sky

Offerings were made at a place named Teotiuacan [sic]. And there all the people raised pyramids for the sun and the moon; then they made many small pyramids where offerings were made. And there leaders were elected, wherefore it is called Teotiuacan [city of the gods]. And when the rulers died, they buried them there. Then they built a pyramid over them. The pyramids now stand like small mountains, though made by hand. There is a hollow where they removed the stone to build the pyramids. And they built the pyramids of the sun and the moon very large, just like mountains. It is unbelievable when it is said they are made by hands, but giants still lived there then.[1]

A Spanish chronicler of the Aztecs

Mr Ellicott drew a true meridional line, by celestial observation, which passes through the area intended for the Capitol. This line he crossed by another, running due east and west, which passes through the same area.[2]

An early (1792) description of how Washington D.C. was laid out

According to Spanish chroniclers who visited it during Europe's Renaissance, the ruling Aztecs of Mexico City said that Teotihuacan, built just a few centuries before the time of Christ and already abandoned for several centuries, was the birthplace of the gods. (Recall their story of creation told in the chapter on "The Storyteller's Sky," and see my first epigraph above.) Teotihuacan was not built randomly, the way the medieval cities of Europe evolved. Excavations reveal that the great city's grid structure was carefully pre-planned, and that it was intended to be as much a holy place as it was a center of civic activity. Viewed from various perspectives, its towering pyramids seem to imitate the mountains that ring the valley (Fig. 35), as the chronicler implies.

Seen from above, the rectilinear urban plan of Teotihuacan appears defiantly stamped upon the natural environment, twisted out of line with the lay of the land. (The course of the San Juan River that runs

35 *Teotihuacan, Mexico: The view is from the north, standing on the Pyramid of the Moon, looking along the Street of the Dead. Note the way the Pyramid of the Sun (left of center) seems to imitate the shape of the mountain that lies beyond it. The east–west axis of the city once aligned with the Pleiades via a pair of pecked cross petroglyphs (like the one shown in the inset), which also may have served as calendrical counting devices. Note the count of twenty points along each axis—the numbers of fingers and toes on one who counts time.*

through it was diverted and canalized to conform to the skewed grid, which is oriented 15½° clockwise from the cardinal directions when viewed from above.) Teotihuacan's main axis, the 5-km (3-mile) long Street of the Dead, as the Aztecs who discovered it in the 14th century would later name it, ends at the base of the great mountain on the north, Cerro Gordo. The "street" is really a series of gradually elevated platforms

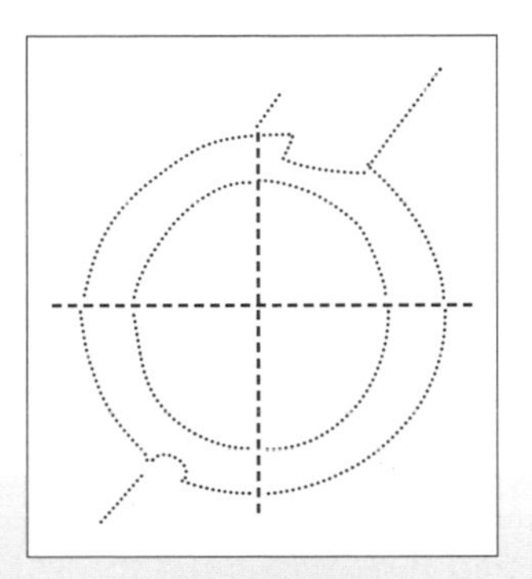

that terminates in the Plaza of the Pyramid of the Moon (shown in the foreground of Fig. 35). At the midway point, and on the east side of the Street of the Dead, lies Teotihuacan's largest structure, the Pyramid of the Sun (it measures 225 m (740 ft) at its base and 75 m (246 ft) in height), its summit positioned due south of its lunar counterpart and in line with the distant mountain, Cerro Patlachique. From its *adosada*, or frontal platform, the east–west street defines a perpendicular axis to the grid that envelops the ethnic *barrios* that make up the perimeter of the 8-square-km (3-square-miles) urban zone. The Ciudadela, a walled compound, is positioned still farther south and is also on the eastern side of the Street of the Dead; its recessed negative space (half a kilometer on a side) seems to balance the upward protruding Pyramid of the Sun. Within that recessed compound lies the Temple of Quetzalcoatl, with its splendidly decorated façade.

More than 2,000 years ago, the builders of Teotihuacan surveyed and laid out one of America's first great cities. It would come to house more than 100,000 people. Along the Street of the Dead, in the stucco floor of a building located just south of the Pyramid of the Sun, lies a clue that serves as visible evidence of the precise course taken by Teotihuacan's architectural planners—a petroglyph pecked into the stucco in the shape of a double circle centered on a cross (Fig. 35, inset). The design closely matches another carved on a rock outcrop 3 km (1.9 miles) to the west of the Sun Pyramid. In the 1960s archaeologists discovered that a line between this pair of architect's benchmarks lies almost exactly parallel to the east–west street of the ancient capital. A third petroglyph on Cerro Gordo, a mountain to the north overlooking the city, along with a fourth on the south, may have marked out other significant geographic directions. Following the method shown in Fig. 25 in Chapter 5, Fig. 36 shows how pairs of sticks might have been used to orient the city.

If you stood over the petroglyph near the Street of the Dead 2,000 years ago, and cast your eye out along the east–west axis of Teotihuacan toward the marker on the western horizon at the correct time of year, you would have seen the Pleiades setting over the mountain horizon. Why align a city with the Pleiades? First, they passed directly overhead in the latitude of Teotihuacan at the time the city was erected, thus signaling the important fifth cardinal direction. And second, the Pleiades' reappearance in the east, after having been lost in the light of the glaring sun for 40 days, happened on the very day the sun also passed the zenith.

Here was a visible, convenient timing mechanism to signal the start of the new year. Tying the sun to the stars, this temporal device was

36 *One possible scheme for achieving celestial alignment of the city of Teotihuacan may have consisted of using a pair of sighting sticks such as these, which are depicted in the later codices. Here hypothetical architects set up an alignment with the rising Pleiades.*

quite different from beginning the solar calendar at one of the solstices, which mark its most northerly or southerly passage, as, for example, at Stonehenge (the Western year begins just after the winter solstice). Being both prominent and in the right place at the right time, the Pleiades would have been the quintessential asterism of choice for the innovative calendar keepers who advised the rulers of Teotihuacan. But the peculiar orientation of Teotihuacan may have been a natural choice because of another curious coincidence. The intervals of time marked out by its axis were ideally tuned to the base-20 body count already in widespread use (see Chapter 10 for a detailed discussion of Mesoamerican timekeeping). The city lay at the precise location where exactly twice the body count in days (i.e. 40 days) also marked the period between sunrise along the axis between the pecked crosses, and the day the sun passed overhead.

Who put this heavenly ordained plan into action, when, and precisely why? What was the sequence of events that led to the planning and arrangement of an entire city according to heavenly dictates? Drawing on the example of the Skidi Pawnee discussed in the previous chapter, we can well imagine that great significance would have been attached to nature's signals at the initiation of the earth's cycle of fertility. The first signs of thunder can be heard in the Valley of Mexico around the time leading up to the first overhead passage of the sun, a phenomenon unique to the tropics (see Chapter 3). Today, around the first of May, dark clouds still gather in the east over Mt Tlaloc, or as the Aztecs called it, *Tlalocan*, home of Tlaloc, the god of rain (recall our discussion of him in Chapter 1). Their appearance heralds the coming of the green-wet season that replaces the brown-dry season in the two-phase Mesoamerican seasonal cycle.

Like the Skidi Pawnee, early settlers at Teotihuacan dating back to the 1st millennium BC may have made some attempts to demonstrate the relationship between their city and the gods who created time. But any evidence of early shrines, appropriately oriented to symbolize this great stress point in the seasonal cycle, has long been erased, for at that time such structures would have been fashioned out of perishable materials. However, recent excavations of the Pyramid of the Moon show that the precise 15°28′ east of north orientation of the east–west axis of the grid was established in stone as early as the 3rd century BC. The earliest phases of the Pyramid of the Moon are misaligned with the later grid. They may represent a first crude attempt to formalize a space-time connection in stone between city and cosmos. And like the Skidi priests who recognized patterns in the stars, clever Teotihuacan skywatchers discovered a brilliant way to set the urban backdrop for conducting their seasonal rituals in harmony with the workings of the cosmos. They chose the most obvious, convenient referents in the sky over their city to signal the time to celebrate the initiation of the planting season. The Pleiades, the setting sun, the body count—all conspired to convey the message that Teotihuacan's grandeur and importance were ordained in the heavens.

Just as the Skidi village rites centered around Tarawa and the creation story, urban Teotihuacan's rituals very likely invoked some form of the pan-Mesoamerican origin myth of multiple creations, the most recent one having resulted from the sacrifice of the gods. The performance of a sacrificial act before the community of the faithful became a literal reenactment of what happened at the time of creation. As we recall from Chapter 1, 1,000 years after the fall of Teotihuacan Aztec legend told of its fifth creation. There all lay in darkness until the bravest of the deities made the supreme sacrifice by throwing himself into the flames of the pre-dawn fire that would become the sun. Thus the mandate: "Let us die so that the sun may be revived" was passed on to the citizens of the Aztec empire, self-appointed inheritors of Teotihuacan's great tradition.

We can be sure that human sacrifice was practiced at Teotihuacan. For example, the Temple of Quetzalcoatl, a pyramid dedicated to the myth of the origin of time, features a grand façade displaying sculptures closely tied to the gods of nature. Effigies of the Great Goddess of the storm, and symbols of the world above (avian creatures) and the world below (the earthbound snake), joined together in the feathered serpent deity, Quetzalcoatl, creator of the divisions of the calendar.

Tunneling into the three sides of the pyramid in the late 1980s, Mexican archaeologists penetrated one burial after another, yielding up a total of more than 200 male victims of sacrifice. Clad in warrior costumes and goggle-eyed masks associated with the storm goddess, each had their arms bound at the back. In a curious juxtaposition, the dead warriors were laid out in straight lines in groups of 9, 13, and 26—all significant numbers that define the structure of time. The sacrificial act paid reverence to time itself.[3]

Still more recently, sacrificial remains have been unearthed at the Pyramid of the Moon.[4] There Japanese archaeologists also found bound sacrificed warriors (probably captives) along with rich offerings consisting of pyrite mirrors, obsidian blades, and the remains of falcons, jaguars, and wolves. To judge from the traces of wooden cages that once contained them, these animals had been placed in the tomb while still alive. Along the central south–north axis of the Street of the Dead lay four bound victims. They were aligned in parallel burials in an east–west direction. An adjacent burial held eighteen decapitated crania. Clearly, the city of Teotihuacan conducted rituals of sacrifice rich with celestial symbolism as a part of the debt payment to their gods. Pleas for assurance of a fertile maize crop probably accompanied such rites.

Lacking both a written account of the rites and any outside witnesses to the events, we can only speculate about how the priests of Teotihuacan set the ritual stage. We can only guess how they told their story of creation, and the words they used in their precisely timed rites to certify that their city would continue to thrive, conducted in the divinely ordained space that they had so carefully constructed and

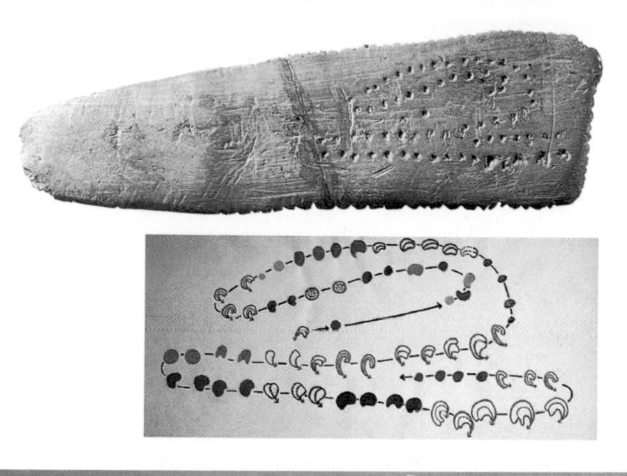

1 This carved piece of bone from the wing of an eagle found in a cave in western France, is dated to approximately 30,000 BC. Were the marks made by a tool sharpener or are they the first notational record of the phase cycle of the moon? The diagram below shows the likely chronological course of the markings and the direction of the turn of the tool used to make them (see p. 76).

2 (*Right*) Carved rock face, Presa de la Mula, Mexico. It could be an ancient bull tally, but more likely represents a 2,000-year-old calendar made by hunter-gatherers, who charted a series of lunar phases to mark out the schedule of their activities (see p. 80).

3 (*Left*) The Batammaliba of Togo and Benin align the crossbeams of their houses so that they point in the direction of the equinox sunrise and sunset. They believe the sun is human in form and lives in his house in the western sky. The doorway of the sun's house faces east; therefore all Batammaliba shrines must open to the west to face him (see p. 119).

4 This section of an 8th-century AD mural painting from the Maya ruins of Bonampak shows victims of conquest supplicating before a warrior chief. Zodiacal signs at the top inform the scene (see p. 168).

5 Maya scribes, possibly astrologers, shown writing in their codices. Elder scribes (far left and fourth from the left), brushes in hand, instruct their students, who seem to exhibit varying degrees of attentiveness. Note the glyphic utterances and the dot-and-bar numbers emanating from both the mouths of the teachers, as well as from the tips of their brushes (see p. 187).

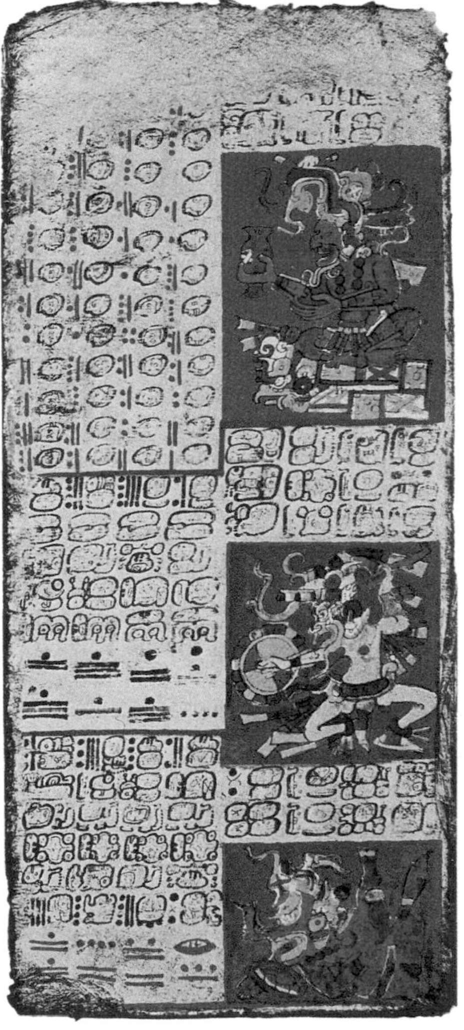

6 (*Above*) These pages of the Dresden Codex constitute part of a Maya Astronomical Table designed to chart the course of the planet Venus. The four dot-and-bar intervals discussed in the text appear across the bottom of the right page. Each eight-day interval (represented by a zero above a bar and three dots) preceding morning heliacal rise is followed by a (middle) picture showing the male Venus deity flinging spears (omens), which represent the dazzling rays of light that accompany his morning star apparition, at victims pictured in the lower picture. An offering to the deity is made in the top picture. The left page is a user's manual consisting of a multiplication table and entry dates (see p. 204).

7 Here the lunar rabbit appears in the Codex Borgia, a *c.* mid-16th-century divinatory document from highland Mexico (see p. 17).

8 Tezcatlipoca, god of the smoking mirror. On his right foot he carries the mirror that he uses to look at the stars and see into the future. Grotesque looking deities like this one frightened the Spanish priests sent to convert the natives—one reason why so many Mesoamerican books were put to the torch (see p. 15).

9 The Gregorian Calendar Reform. As Pope Gregory XIII looks on at the far left, a member of his commission on the reform of the calendar graphically points out the accumulated backslide between the calendar and the seasons, each represented by day markings along the zodiac (see p. 201).

aligned. If bringing rain and crop fertility had anything to do with the ceremonies, they would no doubt have started with a first appearance of the Pleiades. Perhaps the duration of the rites was timed by sunsets between the alignment of the great luminary with the Teotihuacan grid up to the day it reached its overhead point. (We know that the Aztecs conducted ceremonies that lasted twenty days or more.)

Imagine this scenario: the plaza in front of the Pyramid of the Moon is fully illuminated, and the sacrificial victims are paraded up the Street of the Dead. We can see them shackled together as they ascend the great pyramid at its terminus. At the summit they breathe their last breath. They are sacrificed and then entombed. Their remains, which have lain untouched for 2,000 years beneath the pyramids, are our only link to such an imagined past.

So influential were the traditions of the great city that the skewed Teotihuacan grid plan was copied all over Mesoamerica for generations after the city's sudden and inexplicable fall in the 6th century AD. Cities such as Tenayuca in the Valley of Mexico and far away Copan (in Honduras) also have their high mountain on the north, just like Teotihuacan. And dozens of petroglyphs resembling the Teotihuacan pecked crosses have been discovered by archaeologists, at sites ranging from the far north of present-day Mexico to the remote southerly Maya ruins in the Guatemalan rain forest. Images related to war and sacrifice are reflected in the iconography on buildings at many of these sites. For example, the Maya habit of conducting "star wars" and raids based on the favorable positions of celestial bodies, especially Venus (see Chapter 10), likely originated in the city of Teotihuacan. Indeed the Aztec reverence for the great city, home of the gods and the place where time was born, is as justified as our gratitude to Rome and Athens for the gifts that emanate from our own Western Classical tradition.

The habit of orienting entire cities astronomically became widespread throughout ancient Mesoamerica. Even the conquerors who crossed the ocean from 16th-century Spain had heard about it when they entered the Aztec capital of Tenochtitlan—modern Mexico City. One informant told the Spanish friar Toribio Benavente that a festival took place when the sun at the equinox stood in the middle of the great temple, and because the alignment was a little crooked, the emperor Moctezuma needed to pull the temple down and straighten it.

The building in question is the Templo Mayor, or Great Temple, the largest and most centrally located Aztec building in ancient Mexico City. It was thoroughly re-excavated in the early 1980s after electrical workers engaged in installing one of the city's new subway lines accidentally broke into an offertory cache of jades, decorative shells, skulls, and flint knives that had been deposited there 500 years before. Once each of its seven façades was exposed, measurements of the alignments revealed that the orientation exactly permits the rising equinox sun to fall into the notch between the twin temples that once surmounted the flat-topped 31-m (102-ft) high pyramid (Fig. 37).

Spanish chroniclers tell us that a royal observer situated in the plaza below carefully watched for the sun, and when it arrived there the town crier would signal the time to begin the first of the rituals of sacrifice that attended each of the eighteen months of the year. A Spanish priest who chronicled the history of Aztec Mexico City in a twelve-volume work tells what happened in one of the rituals:

> There was the paying of the debt [to the Tlalocs] everywhere on the mountain tops ... And there they left children known as "human paper streamers," those who had two cowlicks on their hair, whose day signs were favorable.[5]

Initiated by sunrise over the Templo Mayor, the ancient Aztec rite of spring set the stage for the payment of the debt to Tlaloc, the rain-bringer—the same fertility god the ancient people of Teotihuacan had worshipped nearly a millennium earlier. This was the time of year to watch for the clouds to begin to form over Tlalocan, the mountain

37 The Aztec Templo Mayor, depicted in a model. Though built nearly 1,500 years after Teotihuacan, it also follows celestial dictates. One early historian tells us that the emperor Moctezuma built it so that, as seen from the plaza below, the sun on the equinox would rise in the space between the twin shrines at its summit. This signaled the proper time for "debt payment" (human sacrifice) to the god of rain and fertility. Inset: a cache of skulls excavated from deep within the temple.

home of the rain god Tlaloc, which also lines up with the axis of the Templo Mayor. Today Mt Tlaloc is still the place where the dark clouds that bring the summer rains begin to form. So there is logic in the Aztec myth that all water comes from the inside of the great hill where the rain god resides. What more compelling offering to the rain god during the youthful part of the year than the precious tears shed by the youngest members of Aztec society? Remains associated with the ancient rain ritual still stand at the mountain's summit. A 300-m (1,000-ft) long causeway built of rough-cut stone leads to a ceremonial precinct where the Nahuatl-speaking descendants of the Aztecs still come to worship a Christianized version of their ancient deity. Thus the alignment of the great city of the Aztecs began in a dialogue with the gods who live in all the sky-earth.

Discourse with sky gods also influenced the structure of the Etruscan cities of ancient Italy. Little written evidence survives, but a curious metal artifact does reveal clues to the plot of a fascinating urban narrative. In 1877 a farmer plowing his fields near the northern Italian town of Ciavarnesco, near Piacenza in Tuscany, unearthed a half-kilogram hemispherical chunk of bronze about 7 × 12 cm (2.8 × 4.7 in) in size. The palm-sized artifact proved to be a one-to-one model of the liver of a sheep, replete with gall bladder and other appendages (Fig. 38). Additional clues from Roman historians led to the conclusion that the liver was an element employed in the Etruscan discipline of divining the will of the gods by examining entrails, a technique derived from their Middle Eastern origins. Likely descended from seafaring people who lived along the Aegean or Anatolian coast, the Etruscans had established a host of cities in the central western Italian peninsula between the 8th and 4th centuries BC. Known for their elaborately decorated multi-chambered tombs

38 *A model of the bronze liver of Piacenza (above) shows the subdivision of the sky according to Etruscan principles of divination. The names on the outer circle of the diagram at the left are based on descriptions of Martianus Capella, a later Roman writer, while those carved on the bronze liver itself list the sixteen sectors of heaven. One way the* haruspex, *or diviner, might have held the bronze liver while instructing or divining is indicated at the lower left. The anatomically correct model includes visible parts of the liver—even the gall bladder.*

that housed huge stone sarcophagi with lids decorated by life-size sculptures of family members, by the mid-6th century BC the Etruscans had expanded their influence as far north as the Veneto region of the north Adriatic coast, and to nascent Rome in the south; most of their cities lay in the region of Tuscany (the Italian province later named for them).

Inscribed on the outer fringe of the liver, in the only partially decipherable Etruscan language, are the names of the sixteen regions of the sky, each corresponding to the name of the Etruscan god believed to reside there. Curiously, the names are arranged hierarchically in clockwise order; they seem to progress from positive toward negatively aspected deities. Thus, in the Etruscan system, the most powerful, fortunate gods, such as the sky deities Tin Cilen (Jupiter) and Uni (Juno), were believed to reside in the sky above the northeastern horizon; those associated with lesser good fortune (the sea and solar deities, such as Catha, daughter of the sun) and lesser ill fortune (earth deities such as Fufluns (Bacchus), god of wine, and Selva (Silvanus), god of the forest) were positioned, respectively, in the southeast and southwest. The most powerful ill-fortuned or infernal deities, such as Veiovis or Vetis (Pluto), god of night, controlled the region of the northwest. Whether it had to do with exacting a simple omen or building a complex urban place, getting the cosmic directions right was a major element in the Etruscan dialogue with gods. The planning of an Etruscan city was a rigorous process therefore, that could be conducted only by trained professionals.

Roman historians tell us that an elaborate rite accompanied the Etruscan process of deciding where to build a city. To begin with, it must always be in a high place, so they would choose the appropriate hilltop by setting out sheep to graze on its flanks. Once several weeks had passed, the *haruspex*, or inspector of entrails, would preside over the sacrifice of the sheep. Likely consulting a liver model, like the one depicted in Fig. 38, he would "read" the extracted livers.

The rationale for such action is really not so far-fetched. For example, the Roman writer Vitruvius tells us that, for reasons of health, it would have been important to examine all forms of animal and vegetable life in the vicinity of a prospective site for any abnormalities that might come from prolonged and possibly adverse dietary dependence on local flora. Unfortunately, because of a lack of textual information (only a handful of Etruscan documents survive), we have no clue regarding how the bronze liver was actually used in practice. It may have been an actual part of the divinatory process, perhaps a handbook of indications gleaned by the *haruspex*, or more likely a teaching device employed in a school for diviners. Once the site was determined, one black and one white ox were yoked and two farmer-priests plowed furrows about the perimeter, in opposite directions, to circumscribe the city's intended space (see p. 120 for a similar Batammaliba rite).

Why the liver? The Etruscans believed in a one-to-one correspondence between the macroscopic structure of the world outside and the microscopic organization of its component parts that lie within us—an idea that resonates with the 260-day calendar in Mesoamerica, which originated in both celestial and bodily cycles, as we will see in Chapter 10. Thus the liver was considered a *templum*, whence the Roman (and hence our own word) "temple."

Modern Western society views the brain as the center of all human thought and feeling, but just a few centuries ago Renaissance scholars believed the heart played the primary role. The Incas divined by examining the inflated lungs of a sacrificed llama, while the Etruscans chose the more conspicuous liver, where all the internal bodily fluids from blood to bile converge. They believed that the *haruspex*, who gazed into the shining essence of the liver, could peer directly into the minds of gods. We may wonder: are those modern cosmologists, who claim to read the mind of God when they probe the initial ripples of the Big Bang creation with powerful telescopes and mathematical formulae, so different?[6]

As at Teotihuacan, archaeology offers us some insight into the Etruscan preoccupation with getting the orientations correct. All Etruscan temples face the southern half of the sky, the region of the lesser gods—the ones most clearly identified with tangible, terrestrial affairs, and consequently perhaps most subject to persuasion by mortals. In those instances where we know the name of the deity to which a temple is dedicated, the structure does seem to line up with the general region of the sky occupied by the deity.[7]

To judge by the archaeological evidence, the Romans, who in a 4th-century BC burst of expansion cruelly decimated the culture that had so long been a thorn in their side, could not care less which way to orient their temples. Their axes spread out over all directions of the compass. However, although they may have lost the more intimate relationship with the heavens possessed by their cultural predecessor, the Romans did acquire the Etruscan habit of orienting their towns on a grid—a habit they have passed down to us.

In founding new cities or Roman colonies, *gromatici* (surveyors) emphatically refer to upholding the ancient Etruscan practice of astronomically orienting towns. For example, the 1st-century AD historian Hyginus Gromaticus says:

> The *limits* were established not without consideration for the celestial system, since the *decumani* [the east–west streets] were laid out according to the sun and the *cardines* [the north–south streets] according to the celestial axis. This system of measurement for the first time established the teachings of the Etruscans; these indeed divided the earth into two parts according to the course of the sun. The part situated to the north they called *right* and that situated to the south they considered *left*; from east to west because the sun and moon are directed in these ways. The other line led from south to north and the parts on the far side of this line they called *antica* and the parts on this side they called *postica*. And from these terms the boundaries of the temples also came to be described.[8]

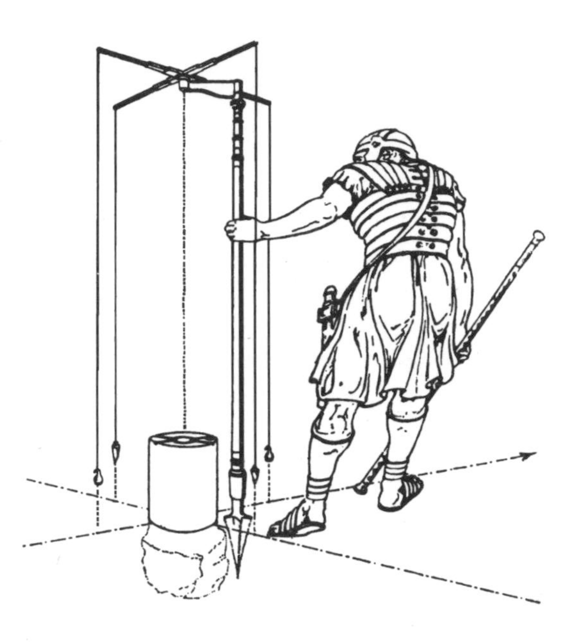

39 *Though lost by Roman times, the complex Etruscan process for laying out a city in accord with the sky gods was passed on to the Romans in the concept of grid structure, known as* centuriation, *or the division of the city into 100 squares. Here the process is initiated by a professional surveyor. The Romans in turn bequeathed it to us.*

But how to accomplish this feat in practice? Hyginus explains: Once the augur (diviner) had left the scene, the engineer would use the mid-point of the sun's appearance to establish the east–west line. First he would draw a circle on the ground, then put a gnomon in the middle. By marking where the shadow entered and left the circle, and drawing a line between these points, he would trace the east–west *decumanus* (see p. 72). The bisector of this line would yield the north–south *cardo*. The famed 1st-century BC Roman architect Vitruvius mentions the same method. In the older, less practical tradition, he tells us that the lateral axis had its two extremes (the solstice points) in the domain of the rising and setting sun. Prolonging the shortest shadow cast by a person would offer a simpler approximation first to get the *cardo*; then this would be bisected to lay out the *decumanus* (Fig. 39).

While the grid-like layout of both Teotihuacan and the Etruscan city seem to have something in common with a bird's-eye view of New York, Moscow, or Tokyo, the lack of a written historical record in many ancient cities keeps us from understanding the details behind the motive for cosmically aligning an urban site. However, cities with long written histories—like Beijing—do provide us with some unanticipated connections.

A mandate from heaven underlay all Chinese dynastic ideology. The written legacy helps us to understand the reasons behind the desire to orient their capital city to the stars. Observational records trace the idea to the strong bond that existed between astrology and good government.

Chinese society has always been bureaucratically organized. Family histories, such as those of the Chin Shu dynasty (*c.* AD 265) contain lengthy chapters on astronomy. These consist mostly of astronomical records, such as where and when celestial objects appeared or disappeared, their color, brightness, direction of motion, and particularly the gathering together of objects in one place. They also include implications, based on past history, that such data might have on family affairs. For example, one Chinese astrologer writes that the Zhou dynasty flourished when the planets, particularly Jupiter, gathered in Roon (Scorpio). According to one analysis of court records, the Zhou launched a military campaign in late autumn of 1047 BC because of Jupiter's favorable aspect.[9] Next time you update your family picture album, imagine needing to take note of where Jupiter is positioned, or how brightly Arcturus glitters!

A merchant trade class like that of medieval Europe did not develop in China. Rather it was the landed gentry of the early feudal class system that charted the direction of star and state. Unlike Plato and Aristotle, who taught in a democratic city-state, Chinese philosophers were intellectuals of the court, and the virtues they instilled in them bonded the agrarian peasant class to the ruling warlord and prince. There was little interstate scholarly communication; so science in China was not as basically outwardly directed as it has been in the West (for a discussion of this topic, see Chapter 11).

The Chinese called their constellations the "heavenly minions." But when they looked to the north they saw not a pair of wheeling bears flanked by a dragon, as we do, but rather a celestial empire. Which constellations did they recognize and what do the Chinese stars tell us about their ideas concerning rulership and the orientation of the city? Confucius, the 6th-century BC Chinese philosopher, compared the emperor's rule with Polaris. Just as the emperor was the axis of the earthly state, his celestial pivot was the Polar constellation. The agrarian economy revolved around the fixed emperor the way all the stars turn about the immovable pole. According to one legend the Divine King was born out of the light radiated upon his mother by the Pole Star. Four of our seven Little Dipper stars plus two others constituted the *Kou Chen*, or "Angular Arranger" of the Chin Shu dynasty. These stars made up the great Purple Palace, and each of their celestial functionaries had its terrestrial social counterpart. One member of the group was the Crown Prince who governed the moon, while another, the Great Emperor, ruled the sun. A third, Son of the Imperial Concubine, governed the Five Planets, while a fourth was the Empress and a fifth the heavenly Palace itself. When the Emperor's star lost its brightness, his earthly counterpart would sacrifice his authority, while the Crown Prince would become anxious when his star appeared dim, especially when it lay to the right of the Emperor.

The four stars that surround the palace star are *Pei Chi*, the "Four Supporters." On the Chinese star maps they appear well situated to perform their task, which is to issue orders to the rest of the state. The Golden Canopy is another constellation made up of seven stars, most of them corresponding to the wrap-around pole-centered stars of our constellation of Draco the dragon. It covered the palatial inhabitants and emissaries. Beyond them lay the conspicuous stars of *Pei Tou*, the Northern Dipper. More concerned with realizing celestial principles here in the earthly realm, these "Seven Regulators" are aptly situated so that they possess the maneuverability to come down close to earth, so that they can inspect the four quarters of the empire. According to one version, the Northern Dipper (Fig. 40) is the carriage of the great theocrat who periodically wheels around the central palace to check things out. Its stars are the source of the *yin* and *yang*, the celebrated two-fold way of knowing that resolves the tension between opposing polarities—male and female, light and dark, active and passive. *Yin* and *yang* wax and wane with cosmic time; they make up the potentiality of the human condition. For every affair of state the starry winds of good and bad fortune blow across the sky.

Why did the rulers adopt the stars of the north? The stars are eternally visible in the sky, never obscured by the horizon, just as the power invested in royality is everlasting. (Recall from Chapter 3 that in

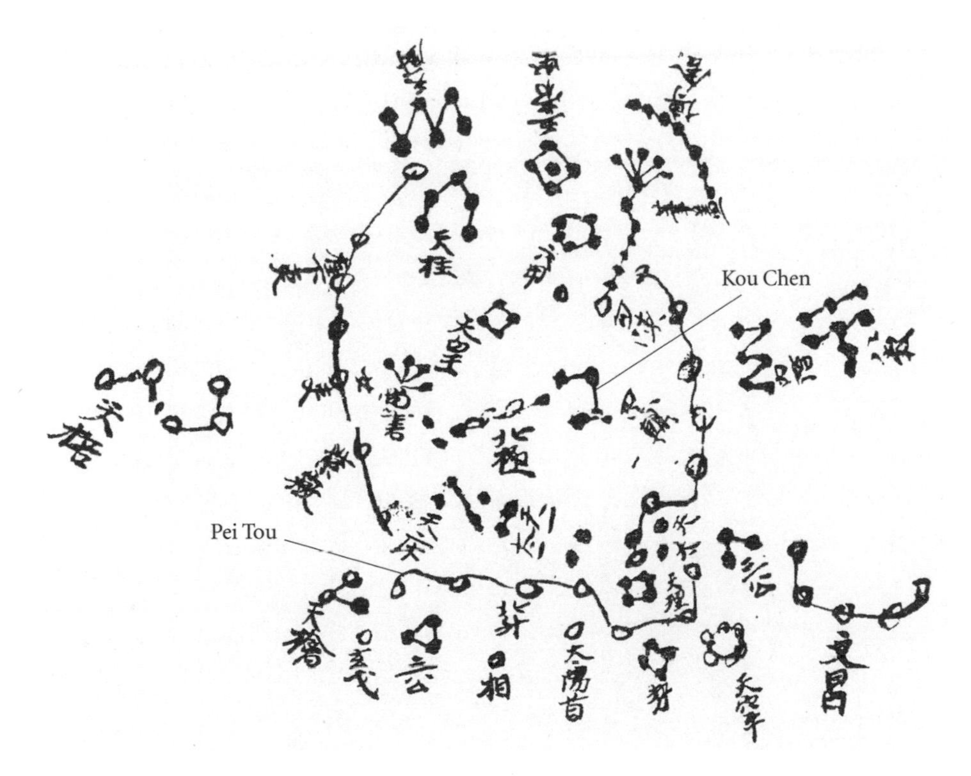

40 *In this star map, dated* AD *940, the polar projection shows the constellations of Beijing's Purple Palace. The Northern Dipper,* Pei Tou, *is easily identifiable. So are four of the stars of the* Kou Chen, *at the center.*

the higher temperate latitudes the stars that turn about the pole are all raised quite high in the sky.) The fixity of the polar axis becomes a universally visible metaphor for the power of the state that developed in these places. (It is no surprise that Cassiopeia the Queen and Cepheus the King of old Western star lore are positioned so near the pole.)

Given the close parallel between the events surrounding the palace economy and the divinely ordained celestial arrangement, it seems logical to inquire whether Chinese royal architecture is also situated so that it lies in perfect harmony with the land- and skyscape.

To harmonize the arrangement of the royal capital with the local contours of cosmic energy, the king would call in his geomancer to perform the art of *feng shui*. Like the Etruscan *haruspex*, this expert would decide precisely where to select a site and how to arrange it; he might also consult the oracle bones. Divining from Chinese oracle bones, widely practiced from 1400 to 200 BC, involved phrasing a question and then piercing a large segment of animal bone—usually the scapula of an ox, deer, pig, or horse—with a heated poker. An interpreter was then consulted to read patterns in the cracks that resulted from the expansion of the heated bone. His sources of cosmic knowledge were the local magnetic field, the paths of streams in and out of the immediate environment, and the categories of land forms. His secretaries would follow him around and take meticulous notes as he performed his task. Sometimes workers would need to remove vast quantities of boulders or plant forests of trees to regulate *yin* and *yang* energies that would pass in and out of the urban area.

A historical account of such an urban foundation ritual survives. It is associated with the city of Lo-yang of the Chou dynasty of north China at the close of the 2nd millennium BC. On the second day of the third month:

> Tiog-Kung (Chou-Kung: the Duke of Chou) began [to lay] the foundations and establish a new [and] important city at Glak (Lo) in the

> eastern state. The people of the four quarters concurred strongly and assembled [for the corvee] ... In the second month, the third quarter, on the sixth day *i-wei* in the morning the King walked from [the capital of] Tiog (Chou) and so reached P'iong (Feng). The Great Protector preceded Diog-Kung to inspect the site. When it came to the third month ... On the third day, *mou-shen*, the Great Protector arrived at Glak in the morning and took the [tortoise] oracle [as bearing] on the site. When he had obtained the oracle, he planned and laid out [the city]. On the third day, *keng hsu*, the Great Protector and all the people of Ien (Yin) began work on the [public] emplacements in the loop of the Glak [river] ...[10]

Note the attention to detail regarding place and time in this lengthy statement. It suggests to us that acquiring proper urban form depended on getting things right—especially the cardinal axes. This was one important task of the *feng shui* expert, for if it were to function properly, the city needed to be accurately partitioned into its quarters.

Today Beijing still preserves its ancient cosmic plan. If you stand in Tiananmen Square you can line up the Bell and Drum Towers, the Monument to the People's Heroes, and the Mausoleum of Mao Zedong on a perfect north–south axis. Continue that line and you will discover that it runs through the gates of the old city. Today the cosmic axis is defined by a marble sidewalk that marks the imperial meridian. The Hall of Supreme Harmony, which houses the emperor's throne, lies at its northern terminus; this symbolizes the fixed circumpolar region, the place, as the ancient texts say, where the earth meets the sky.[11]

Beijing offers a lasting reminder of the cosmically ordained schedule of duties of the emperor. Like his Aztec counterpart, he must perform a specific task at the beginning of the first month of each of the seasons, these being determined by the court astronomers who followed the course of the moon and sun as well as the five planets across the lunar mansions of the Chinese zodiac (see Chapter 2). For example, the emperor would go to the eastern quarter of his domain to start the new year every spring equinox. He would pray for a sound harvest; then, followed by his ministers, he would plow a ceremonial furrow in one of the fields. At the other seasonal pivots he would visit the other quarters of his city so that everyone with a place in his domain would be accorded their own special time.

This formal quadripartite seasonal calendar would have been familiar to any farmer, for it was based on what he could clearly see in the sky. At the beginning of summer Antares lay due south at sunset, while on the first of winter the Tristar of Orion's Belt took its place. Of course, farmers knew well when they could plant, but they needed to be supremely aware that the *official* time to do so occurred when the handle of the Dipper pointed straight down; for then and only then was it the first day of spring—the time for the king to come forth and speak to the people about the new year's harvest.

In its high-level form, the keeping of the observations and the preparation of the calendar resided in the state observatory. This institution lay well hidden within the bowels of the Purple Palace. The importance of astronomical observing in the world of politics made such secrecy, even in the closest quarters, a necessity. One directive issued by a 9th-century AD Tang Dynasty king reads:

> If we hear of any intercourse between the astronomical officials or of their subordinates, and officials of any other government departments, or miscellaneous common people, it will be regarded as a violation of security relations which should be strictly adhered to. From now onwards, therefore, the astronomical officials are on no account to mix with civil servants and common people in general. Let the Censorate see to it.[12]

And so the astronomers, spurred on by their government, dutifully performed their appointed task: to give the correct time so that the affairs of state might be properly conducted.

In this chapter we have discovered how archaeology and history together provide strong evidence about how and why ancient people established a dialogue between city and sky. Skidi village, Mesoamerican metropolis, Etruscan city, ancient Chinese capital—their pre-planned designs all share a basic motive: providing a backdrop for the great conversation with the transcendent. That earthly stage was first the village, then the city. The set design seems strange to us, a stage upon which deities reside in the heavenly realm reflected in the unerring, unceasing movement of the sun, moon, planets, and stars across a pristine sky, the whole of it pivoted about a fixed axis that defines distinct celestial domains and their sacred divisions. As one historian aptly put it:

> To live in the city was simultaneously to have all cosmological knowledge presented to the senses, for the city, whether Rome with its cardo or Tenochtitlan with its equinoctial axis, paralleled the cosmos in its layout, in its orientation ... in the way it transmitted the continuity between the body and the world ...[13]

Skidi chief, Estruscan *haruspex*, Teotihuacan and Tenochtitlan calendar keeper, Chinese *feng shui* expert—each used their professional knowledge of the natural environment to dare to read the minds of the gods. These wise specialists were charged with the awesome responsibility of seeing to it that the places established here below would offer the most suitable human entry into the realm of the future—a future gleaned through celestial omens decipherable only in a sacred environment patterned after heaven itself. Beneath the structure of practically every city we find not only mere motives of convenience, accessibility of resources, orientation to favorable winds and sunlight, protection from harm, but also a shared communal *ideology*. And if we dig deep enough we may be surprised to discover that many of our own cities are no exception.

My second epigraph is about the cosmically mandated arrangement of Washington, D.C. The technique of drawing a line from the median position of Polaris perpendicular to the sun's daily path and extending it to the south (as implied in the quote I chose) is exactly the method used to construct ancient Beijing. Washington is the quintessentially planned American city, a product of the French Enlightenment idea of what a great city ought to be. Conceived by Pierre Charles L'Enfant, its grandiose design was intended to rival the great cities of Europe.[14] Washington is laid out in the form of a perfect square, ten miles on a side (the number of fingers on the hands of each of its inhabitants), with its points precisely aligned to the cardinal directions, like so many great cities of the ancient world. The U.S. Capitol (which is accessible by 365 steps, one for each day of the year) replaces the pyramid or *ziggurat* that stood at the center.

Whether monarchy or democracy, every government must establish its connection with the gods, wrote Rousseau. And so, the city of Washington's spoked wheel, with an overlying rectangular grid pattern (terribly confusing for pedestrians as well as for drivers) evokes in the pilgrim—today's tourist—a sense of the national mythology. For example, the city's processional way from the Capitol to the White House along Pennsylvania Avenue, and its inviting line of sight from the Washington Monument along the Mall to the Capitol, mimic the ancient custom of laying out ceremonial ways and pilgrimage routes, such as Teotihuacan's Street of the Dead. Washington presents itself as a center of power, a junction point of sacred and secular space conceived in the Age of Enlightenment—a place that imitates the geometry of the universe and the harmony of the worlds. As ancient Beijing was to the Chinese and Tenochtitlan to the Aztecs, Washington is every American's city in the sky.

CHAPTER 8

The Ruler's Sky

Perhaps there were never any gods without kings, or kings without gods.[1]

A 19th-century British historian

On the morning of 22 March 1622, Opechancanough, young chief of the Powhatan Native North Americans, coordinated a brilliant attack by more than a dozen allied tribes on English colonial settlements along the James River, situated in what today we call Virginia. With pinpoint precision, groups of warriors, each assigned their own village, struck plantations. They burned houses and barns, slaughtered men, women, children, and livestock, and mutilated the remains—all in opposition to two decades of encroachment by the white invaders. Although the attack was successful, wiping out nearly a fifth of a colony that contained more than a thousand white settlers, the foreigners continued to populate and expand the coastal settlements, relentlessly seizing the land and murdering any natives who had occupied it, regarding them as little more than savages—"naked, tanned, deformed ... no other than wild beasts," and "worse than their old Deuill which they worship,"[2] as one colonist put it. This prompted a second, even more effective simultaneous offensive of the allied tribes on 18 April 1644,[3] led by an even more seasoned Opechancanough, descendant of the legendary Chief Powhatan, who had organized the original alliance (Fig. 41).

41 *Powhatan's attack on colonial Jamestown as imaginatively portrayed by Theodore de Bry.*

Both attacks took place in early spring,[4] when the Indians knew the English were low on supplies, a result of both the harsh winter and the absence of any ships from abroad in the intervening months of unfavorable sea travel, and during which they had depleted what remained of the harvest. Early spring was also a time when the tribes were hunting and gathering, not yet having resettled back on the land, where they would be subject to quick retaliation by the enemy. Historians have wondered: how could Opechancanough have coordinated such successful war plans? How did he manage to execute several simultaneous raids so precisely?

Wrote one Jamestown colonist and historian:

> When they intend any warrs ... a Weroance of some lusty fellow is appointed Captayne over a Nation or Regiment to be led forth, and when they would presse a number of Soldiers to be ready by a daie, an officer is dispatcht away, who coming into the Townes, or otherwise

> meeting such whome he hath order to warne, he strykes them over the back a sound blow with a bastinado, and byds them be ready to serve the great king, and tells them the Randivous, from whence they dare not at the tyme appointed be absent.[5]

The Weroance, a messenger-shaman of Chief Opechancanough, could have run or traveled by canoe to set the day, and then direct each warrior group to the proximate English settlement. Since the allied tribes were not much farther than a brief day trip from one another, one would think that little planning would be necessary to surprise the English, hitting all the villages at sunrise on the same morning so that any one would not be given the opportunity to warn its neighbor. But making war requires extensive preparation: you need to stock up on weapons, case your quarry. An account of the first uprising says, "Several days before this bloodthirsty people put their plan into execution they led some of our people through very dangerous woods," and, "On Friday before the day appointed by them for the attack they visited, entirely unarmed, some of our people in their dwellings."[6] One factor often overlooked in the study of indigenous military conduct is that the appropriate war rituals must be practiced prior to an attack. But there had to be a way for Openchancanough to synchronize the "day appointed"—an appointment from which none of his allies could afford to be absent.

I believe the Jamestown raids might more appropriately be termed "Moon Wars," for there is good evidence that the moon was consulted both as a timing device and a symbol behind this particular act of war. As we learned in Chapter 4, hunter-gatherers and semi-sedentary people had full control of the seasonal calendar. The Powhatans and their neighbors were no exception. Some were quite precise when it came to lunar reckoning. For example, we know that the Delaware named the phases of the same moon: the new moon (likely the first visible crescent for them), the round or full moon; and the half round (probably last quarter) moon. The intervals between the directly visible phases—ranging from a few to several days—proved convenient in practical day-to-day operations. Think of how often you refer to activities that will take place "after the weekend," "early next week," or "in two weeks." (Captain John Smith tells us that the Powhatans counted small numbers of days on the fingers and that they used knotted strings and notched sticks to tabulate longer intervals.)

"Little Corn," "Great Corn," "Turkey," "Cold Meal," and "Deer" are a long way from our abstract January, February, March, and April, but what we learn from Native American timekeeping is that the names given to time periods represent "lived time"—the activity itself (see Chapter 4 for other examples of named moons in the lunar calendar). In any dominant society such activity necessarily would include scheduling the conduct of a war.

So, what was happening in Virginia skies at the time of the Powhatan coups of 1622 and 1644? On the eve of the 22 March 1622 attack, the moon was in its third quarter phase; it rose in the south–southeast in the constellation of Sagittarius about an hour past midnight. Interestingly, the face of the moon presented almost precisely the same aspect on the eve of the 18 April 1644 attack: last quarter, rising this time in Capricorn, but in the same direction, also about an hour after midnight. If Opechancanough wanted to plan the attacks on these, or, for that matter, any other dates, the most obvious way to convey his intent to his cohorts would have been to preset the lunar clock by counting days from the first visible crescent. For more effective long-range planning the Powhatans, an association of scattered tribes, could have synchronized the strikes to take place on a particular moon in the cycle, provided they shared an intertribal calendar—not an unlikely assumption. For the 1622 attack, in the latitude of Virginia, the first visible crescent occurred 1 March; for the 1644 episode it happened 28 March—in both cases 21 days, or three-quarters of a lunar cycle prior to the dramatic result that history records.

Why, then, did the two attacks occur a month apart—one in March, the other in April—as reckoned by our calendar? Here we need to call to mind two facts we have already learned about native timekeeping: first, the moon almost always takes precedence over the sun in unwritten calendar systems; and second, there can be twelve or thirteen moon cycles in a lunar year. Suppose, for example, that we count moons from the December solstice, which is 21 December, the day when the sun rises and sets farthest to the south.

Year One ends, then, with the completion of the twelfth moon, eleven days short of the winter solstice of Year Two. The first moon of Year Two will begin with the observation of the first crescent around 10 December. If Year Two also contains twelve moons, then the first moon of Year Three will begin about 30 November. Clearly, the next year cycle, or Year Four, would be a most convenient one in which to insert a thirteenth moon. Now, if we count to last quarter from the first crescent of Moon One of Year One, we arrive at 21 December, plus 21 days, or 11 January. But if we performed the same operation in, say, Year Three, we land on 21 December (30 November, plus 21 days). From this hypothetical example it is easy to see that, with a casual system for intercalating months, identical dates in the Powhatan moon calendar could correspond to dates up to a month apart in our sun-based seasonal calendar.

The historical record conflicts on whether Opechancanough took the Christian holiday of Easter (which he surely knew about) into account when planning the attacks. What makes this difficult to corroborate in real time is that Easter is computed in a wide variety of ways, so that reckoning when it was celebrated is difficult. (There is a touch of irony in the fact that our way of calculating Easter depends on the Hebrew lunar calendar, and, like dates set in the Powhatan lunar calendar, it, too, is a movable holiday—it floats in the seasonal calendar.)

To make matters even more difficult, because the events under consideration occurred in the century following major corrections of the

Western calendar (see Chapter 11), there has been considerable confusion over whether dates reported by Virginia's earliest historians are given in the Julian, or Old Style, calendar in use in Roman Catholic Europe before 1582, or in the Gregorian, New Style, calendar adopted since. As we will learn in Chapter 11, the two calendars differed by eleven days at that time; but the computation of Easter yields dates up to a month apart. Thus, in 1622 Easter Sunday (Old Style) fell on 21 April, but it occurred on 27 March (New Style). Could it be that those who noted that the attack took place around the Easter holiday were referring the Old Style 22 March date to the New Style celebration of Easter, five days before which it took place?

By striking coincidence both versions of Easter fell on exactly the same dates in 1644, the year of the second uprising. But this time if we wish to juxtapose it with the Paschal holiday, as some historical accounts require, we must assume that the record refers to the Old Style Easter date. In that case the attack would have taken place three days before Easter. Unfortunately no almanacs survive from 1644, but an extrapolation of an Anglican common prayer book dated 1641 suggests that the Old Style April date is indeed the most likely one to have been recognized in the colonies.

There is a flip side to the coin of lunar-motivated warfare—symbolism. Though the argument of lunar synchrony is manageable enough, it is also possible that the waning quarter moon possessed symbolic currency among the allied native tribes. The cycle of the waxing and waning moon has long been a universal celestial metaphor for the nature of dynastic rulership (see Box: *The Life Cycle of the Man in the Moon*). As in the myth of the man in the moon, the new ruler is resurrected from the ashes of his father, whose power he acquires by kinship. Scholars agree that the 1622 assault, which had been in the planning stages for months, was also timed to coincide with a funerary ritual in honor of the Great Powhatan, who had died four years earlier. His body had been placed

The Life Cycle of the Man in the Moon

The face of the man in the moon. Compare his face (sketched out here) with the photo in Chapter 1, Fig. 2, where the Aztec rabbit in the moon is sketched alongside.

Employing the qualities of the ancient imagination expressed in "The Storyteller's Sky" (Chapter 1), it is easy to see how the cycle of the phases of the moon, expressed as a narrative featuring the face of the man in the moon, outlined here in profile on the disk, parallels the universal myth of the life-cycle of the heroic warrior. Like that of a young king who accedes to the throne, the man in the moon's career waxes to brilliance and success, his most influential period of rule occurring at the peak of effulgence, when his face is fully filled out. Then the dragon of darkness nibbles away at his countenance, seeking to conquer and destroy him the way he had dethroned the man in the moon's father, the old king. But the moon of the next generation of sky gods emerges out of his father's ashes, to challenge and conquer the forces of evil all over again in an endless cycle.

in a shrine, as was the custom, and allowed to decompose so that the bones could then be taken away for burial. The ceremony attending this last ritual, which symbolized the transfer of power to the new ruler, had taken place only months prior to the first raid on the English. Clearly, young Opechancanough was only beginning to emerge as a leader. And so, the waning quarter moon, riding high in the sky, may have served as a reminder that Opechancanough's authority to command the alliance had been acquired from his revered predecessor.

Opechancanough needed to validate his power at this time in particular. Three of the tribes that lived close to the shore had recently defected to the English side. When the new chief tried to obtain certain herbs from them for the purpose of poisoning the English, the local chief not only refused but also relayed the incident to the English at Jamestown.

We don't know enough about Powhatan cosmic symbolism to speculate much further, save to say the attack was a success. And this is probably why the young chief decided to make the attack 22 years later his second "Moon War." But Opechancanough was not the first ruler to appeal to the sky as a source of power.

In many ancient cultures the ruler was viewed as the embodiment of divine power on earth. Conceived by gods and born of goddesses, the king is the living redeemer, the instrument of divine will. In India, Manu, the supreme lord, was thought to have created the first king when he harnessed the power of the gods of wind, the moon and sun, and fire. Babylonian kings were regarded as direct descendants of the gods:

> When Anu had created the heavens,
> [And] Nudimmud had built the Apsû, his dwelling
> Ea nipped off clay in the Apsû,
> He created Kulla for the restoration of [the temples];
> He created the reed marsh and the forest for the work of [their] construction;...
> He created the mountains and the seas for whatever [...],
> [And] He created the king, for the maintenance of the
> temples;...[7]

For the Babylonians all power resided ultimately in the heavens. By following the course of the planetary deities, the court astrologers could chart the most prudent course of action for the rulers. (We will learn more about how astrology works in Chapter 9.) Thus the king and his retinue remained in close contact and alliance with the forces of nature, each of which displayed to the careful observer his most particular aspect. There was, for example, Nabu the Wise, the Babylonian equivalent of Mercury, who became Rome's swift messenger god because he moved so rapidly across the sky. By keeping his ear close to the ground Nabu developed an extraordinary capacity to read the minds of his fellow deities. Jupiter, the king of gods, who descended from the Babylonian Marduk, occupied the median position both in brightness and in speed. Thus he came to represent the moderation that characterizes those who dole out justice. And fiery red Mars (Nergal in Babylon) was the warlike god of the underworld. His retrograde turnings were particularly influential in ancient China: "When Mars retrogrades in Ying-she, ministers conspire and soldiers revolt,"[8] reads one omen. Ishtar, later Venus, was the goddess of love. They say she had an affair with Shamash, the sun god. She always stayed close to him, hovering over the eastern horizon before he rose or standing high in the west over the place where he set. Often she would disappear into the underworld with him.

The idea that leaders acquire their powers from the forces of nature is recognized the world over. It was a hallmark of the Inca ruler and his followers, who called themselves the Children of Inti (the Sun). An old Inca tale relates how all people once lived in fear of the fire monster Huallalo Caruincho, whose wrath they could temporarily avoid only by making human sacrifices to him—until the hero deity Paria Caca did away with him. Paria Caca was a storm god who took the form of a snow-capped mountain at the top of the world and became human by taking the form of five magic eggs. The eggs hatched five falcons who were then transformed into five men, the heads of the first clans from whom all the Incas descended.[9]

For the Quiché Maya of Guatemala we recall (Chapter 1) that the first clan leaders were four "mother-fathers" who were molded by the divine Bearer-Begetter out of:

> yellow corn, white corn, alone for the flesh,
> food alone for the human legs and arms ...
> It was staples alone that made up our flesh.[10]

When the sun rose he became the source of their power because his unbearable heat was able to turn the enemy deities—animals who would otherwise devour the nascent human race—into stone.

If victory in battle over the enemy of the forces of nature ensures the power of Zeus or Marduk, winning the battle here in the world below the sky endorses the legitimacy of rule of the king. And so, Indra, the Vedic sun king, was said to have triumphed over the demons of darkness, thus fructifying the land.

Hawaiian kings acquired *mana*, or spiritual power, directly from the sky gods via the bloodline, for they say that the deities were their ancestors. The king was regarded as "a god that could be seen."[11] How good his *mana* was depended on performance: how prosperous the harvest; how fat the cattle; how healthy his subjects? As chief mediator between the people and the forces of nature, the king needed to be recharged or reinvested each year during the last month of the dry season. The last such ceremony, recorded in the late 19th century, was timed by the first appearance of the Pleiades in the east after sunset (about 15 November in our calendar). Given the signal from the sky the king entered a special temple, where he offered bananas, coconuts, and a sacrificed pig to the gods of the old year and to feed the months of the new—each timed by the first appearance of its own star, whose name he shouted out.[12]

Egyptian pharaohs reckoned their celestial power source by aligning their pyramids with the pivot of all heavenly motion—the Pole Star. A narrow channel entering the north face of the Great Pyramid tilts upward from horizontal by 31°. It points to the place occupied by the Pole Star in 2700 BC, when the structure was built. This offered the deceased king direct access to his ancestry. His soul could ascend the shaft, pass among the stars, and affect their movements. Like the circumpolar stars of the Chinese emperor mentioned in the previous chapter, the lights of the north, which belong to the pharaoh, are present in perpetuity—just like the ruler.

Three millennia later the early 3rd-century AD eccentric Roman emperor Elagabalus Antoninus (also called Heliogabalus) dared take the name of the sun for his own, a habit that persisted two millennia later when France's Louis XIV insisted, barely a century prior to the Revolution, that French courtiers hail him as "Le Roi Soleil," the Sun King. Elagabalus acquired his name from a Latinization of the Semitic deity El Gabal. He is also responsible for re-popularizing the "Sol Invictus," or day of the "Unconquerable Sun," what we call the winter solstice. Passed down from Zoroastrianism worship of the sun god, Mithra, the power of the ruler is allegedly derived from the "grace" of the creator, a kind of dazzling supernatural aura or glory originating in the sun that surrounded all deities. They alone possessed the capacity to shed this divine light upon deserving young princes who had proven themselves in battle. Once acquired, lest it be abused, Sol Invictus guaranteed perpetual victory.

Among the ancient Maya the king served as the conduit through whom the supernatural powers of the forces of nature manifested in wind, rain, clouds, smoke, water, earth, and sky were passed on to the people. What the pyramid was to pharaoh, so was the stela to the Maya ruler—an embodiment of power that emanated from the sky and carved in stone. These monuments, many of which still stand at the major Maya archaeological sites, depict blood-letting rituals performed by the ruler. He is often shown accessing life's precious liquid, usually from the genitals, which are pierced with a sting-ray spine, the blood dripping onto parchment. The paper is then burned and the smoke allowed to rise toward the ancestor gods. In this act of reciprocity, the king's blood nourishes them, just as they nurture maize and people here in the world below. As we shall see, at the height of the Maya Classic period in the 7th century, rulers commissioned large numbers of carved stelae to herald their success in divinely inspired "star wars."

In the mid-1940s an explorer commissioned by The United Fruit Company stumbled upon the lost Maya ruins of Bonampak deep in the Lacandon rain forest of Chiapas, Mexico. In the interior of its principal temple, and still visible despite centuries of mold and encrustment, he beheld the most exquisite polychrome plaster wall paintings ever seen in that part of the world. The scenes, which belied the notion (held up to that time) of a peaceful Maya race interested only in esoteric contemplation, showed what happened to the losers in battles the Maya engaged. One scene depicting a battle reveals a cadre of jaguar-garbed warriors spearing their enemy; another shows humiliated captives being displayed back at the home site; and in a third mural the victims are displayed in bowed posture, some with blood dripping from their fingers, others already having been decapitated (Plate 4). The survivors seem to be begging their conqueror for mercy.

He is a young heir to the throne, shown elaborately clad in a jaguar pelt and tasseled headdress, just about to participate in the ceremony of his accession to office. A sky panel above mirrors the source of his divine power. The scene displays some of the constellations of the Maya zodiac, each adorned with *chac ek* or "great star" hieroglyphs (the same symbols written in the Venus Table of the Dresden Codex, which we will discuss in Chapter 10).

Perhaps *chac ek* in this case actually is Venus. Dates carved on stelae outside the building that commemorate the events pictured in the wall murals seem to be associated with key positions of the planet. Almost all the dates involved coincide with heliacal risings and settings of Venus, and all of them fall in the November to mid-February portion of the seasonal calendar, a fitting time for war, when the harvest is finished and crops lie ripe in the fields, thus providing sustenance for the invading army. Many of the dates also mark the time when Venus begins its first perceptible descent from the time it stands highest above the horizon. Was this a visual metaphor for the turning point that signaled the battle event? The major battle seems to have been timed to coincide with the first appearance of Venus as morning star (2 August AD 792). This event took place in a region between the constellations of Orion and Gemini, whose corresponding signs are depicted in the sky panel.

The connection between Venus and Maya warfare may have spread well beyond Bonampak during the Classic Maya period. At the ruins of Cacaxtla, atop a huge fortress in the highlands of Mexico far from Yucatan, archaeologists uncovered a set of mural paintings that replicate many of the themes of the Bonampak murals. The battle has just concluded and once again the victors, clad in robes studded with symbols of Venus, stand gloating over their bloodied adversary, who lie broken and crumpled in the wake of their assault. The painting is framed in a border made up of half five-pointed stars, which stand in for Venus in central Mexico. In another room the lords of Cacaxtla, looking as powerful as the radiant deity they worship, are shown beheading the survivors and offering their remains up to the gods of fertility in order to ensure an abundant new maize crop.

During Classic times Maya rulers seem to have adopted aspects of their patron planets the way modern sports fans identify with their favorite team. For example, one of Copan's kings, Yax Pasah, or "First-Sun-At-Horizon" as his hieroglyphic name translates, seems to have favored morning over evening star Venus appearances to schedule his raids on nearby cities. His grandfather, Waxlajuun Ub'aah K'awiil, also called 18-Rabbit after the resemblance of his hieroglyphic name to a figure of that animal, had done the same two generations before him. 18-Rabbit even had built for himself a temple dedicated to Venus. Its façade was adorned with a double-headed sky serpent with a Venus glyph at one end and a sun sign at the other, perhaps mimicking in stone the sinuous movement over time of the imaginary line that connects the sun to Venus in the sky. Copan's Venus temple even sported a special slotted window in its west façade through which court astronomers could carefully time the evening appearances of his patron planet—an astronomical almanac in stone. Modern Yucatec-speaking Maya people have told inquiring anthropologists of an old myth about a serpent with two heads who lives under the world. Every night he positions himself to swallow the descending sun, which his other head proceeds to disgorge at the eastern horizon at dawn.

Monumental inscriptions tied to Yax Pasah's father, K'ahk' Joplaj Chan K'awiil, alias Smoke Monkey, turn out to be a mirror image of those of his father, 18-Rabbit. Smoke Monkey seems to have hitched his affairs to Venus as morning rather than evening star. The dedication of his palace was timed to occur two Venus rounds after that of his predecessor, to coincide with a morning heliacal rise event. Smoke Monkey even saw fit to close off the evening star-facing window in his father's temple. Why would a son revert to the celestial habits of his grandfather? Even kings are no exception to the dictates of human behavior when it comes to kinship. There may be in every offspring an impulse to be different from one's parents. In the serious world of juxtaposing royal Maya history and astronomy, young Smoke Monkey's measure of independence seems to have been expressed in the selection of certain aspects of a celestial body that gave him access to particular powers of the sky deities.

While the Venus cult flourished at Copan in the late 7th century, far to the west in the city of Palenque, the inscriptions of the ruler Kan Bahlam (Jaguar Serpent) suggest that he held a special fondness for the planet Jupiter. Perhaps because he had succeeded a very famous father, Pacal the Great (after the *pacal* (shield) glyph that appears in his hieroglyphic name), who had ruled Palenque for several decades during which he had constructed its most elaborate temples up to that time, young Kan Bahlam needed to distinguish himself. Whatever the reason, the events in his life seem deliberately attached to movements of Jupiter, specifically to the two stationary points (the places where Jupiter changes direction) in its retrograde loop. For example, his heir-designation ceremony, accession to rule, and apotheosis after death all fell within a few days of a second stationary position in Jovian retrograde.[13]

One of the most spectacular of all celestially based events recorded in Maya monumental inscriptions has been attributed to Kan Bahlam of Palenque. It happened on 20 July AD 690, the Maya date called 2 Cib 14 Mol in the 52-year Calendar Round. (The first half of this date is the position in the 260-day cycle, while the second half, the 14th day of the Month of Mol, gives the position in the 365-day year.)[14] Thanks to modern astronomical calculations, we know that on this night a rare planetary conjunction took place in the skies over ancient Palenque. It involved Mars, Jupiter, and Saturn, which had been dancing about the night sky ever more closely together. The date is inscribed in several texts on the stuccoed plaques at Palenque's ruins. Each inscription seems to imply that the three major sky gods responsible for the most recent cyclic creation of the world, which is stated in the inscriptions to have taken place at Palenque some 4,000 years earlier, were re-assembling in the sky. Their conjunction would reaffirm the continuation of the cosmically destined, divine dynasty via Kan Bahlam's ascent to office.

To predict the future, to acquire an intimate foreknowledge of the whereabouts of the sun, moon, planets, and stars that rule day and night, to feel the cosmic power these luminaries display in their various aspects and appearances, and, finally, to embrace the conduct of cosmically derived power in the conduct of human affairs—all of this is virtually impossible for us to fathom. When we point our telescopes toward the moon we witness a world trod upon by astronauts. In our minds stars are blazing infernos powered by high-speed collisions of subatomic nuclei, and our planets are familiar only as worlds of coordinate rank with the one we inhabit. Captivated by our habit of objectifying the universe, we pay the price of never being able to truly understand how ancient rulers acquired power from the sky.

The notion of divine descent of rulership channeled through nature escapes us. Yet we must appreciate that belief in such sanctioned divine rule passed on through the bloodline was once practiced the world over, and for a lot longer than any other form of governance, even if today it has become part of a lost theology. Only the king can perform wonders; only the legitimate heir has access to the gods; only he can forge the link between divine law and moral law. The king embodies justice and, as Plato wrote, "Justice is no other than the sun."[15] Like the sun, justice influences and controls all things, and like the sun it shines constantly, never exceeding its limits.

In a world in full contact with nature, worlds like those of the Maya, the Incas, the Egyptians, the Babylonians, and the Native American tribes of Virginia, it makes perfect sense to celebrate the likeness between the divine rulers in heaven and those here on earth. They are as perpetual as the sun, moon, and stars, who are there for us, all of us—rich and poor, noble and commoner—to contemplate both by night and by day. They are as dependable as the day that follows the night or the spring that succeeds the winter. Believing in a ruler who acquires his power from the forces of nature was as natural to ancient socities as our contemporary belief in a free world in a democratically governed state, whose success is guaranteed by a constitution and the electorate.

CHAPTER 9

The Astrologer's Sky

... hereditary prince and count, sole companion, wise in sacred writings, who observes everything in heaven and earth, clear-eyed in observing the stars, among which there is no erring; who announces rising and setting at their times, with the Gods who foretell the future[1]

Inscription on a statue of an Egyptian astrologer

We all share in the deep desire to know what will happen to us—in a moment, a day, a week, a year, a decade. And history demonstrates that we will use any means at our disposal to peek around the corner of time to read the will of nature revealed through the gods. Sorcerer, diviner, magician, priest, prophet, shaman—each is a term we use to define the role of the specialist who has access to the powers of nature. Only these intermediaries are alleged to possess the special gift of being able to access what is otherwise unknowable, and consequently unsettling to the rest of us.

We call priests adepts in the service of an organized enterprise that is part of a sacred tradition, while we tend to label as prophet, magician, or sorcerer those whose extraordinary powers are unconnected to any recognized religious institution and are mandated by an acquired charisma, often displayed in an ecstatic state. Likewise we differentiate between, on the one hand, the priest, whose power derives from a specific body of ritual knowledge passed down from a predecessor who is deemed legitimate, and, on the other, a shaman, who is divinely gifted and more in direct contact with and consequently able to influence the actions of supernatural forces. The priest performs his rite in public and the symbols connected to that rite can be perceived by an entire congregation. The shaman's rite is a seance in which he reveals his personal visions.[2]

Strictly speaking, divination has to do with inquiring into the future by addressing a specific deity or power through particular media, via some sort of mechanical manipulation. It is a practice the modern world ascribes largely to magicians and shamans—experts who operate outside the confines of society's accepted religious and scientific norms.

I keep a list of the variety of matter and materials employed in the common quest for knowledge of the future throughout world cultures; it reveals the extraordinary lengths to which mediums have stretched their alleged capabilities to acquire such information. I call it my "catalogue of mancies," after the root word *mantic*, which means pertaining to prayer or praying.[3] For example, *geomancy* pays special attention to the flow of magnetic or other energetic currents in the earth to divine the future, *hydromancy* applies to watching the behavior of water, *pyromancy* to fire, and *aeromancy* to air. To the contemporary materialist, for whom predicting the ways of nature is best conducted by controlled scientific experiment (which, too, has its limitations), many of these practices seem quite bizarre, for example, *axinomancy*—divining by a balanced axe, *bibliomancy*—via randomly selected passages in a text, *cledonomancy*—by chance remarks in a conversation, *myomancy*—by observing the movement of mice, *uromancy*—by gazing into puddles of urine, and *gyromancy*—by whirling around until you get dizzy and fall down.

Some mancies have been more widely practiced than others, thus indicating a widespread faith in the best ways to acquire future knowledge. For example, divining by oracle[4] through a trance medium has a long tradition among Mongolian, Siberian, and

Alaskan nomads as well as in Egypt and the Classical world. Divining by oracle bone, as we learned in Chapter 7, was common in China. Like the Inca diviner who read the intestines of a sacrificed llama, or the Etruscan *haruspex* who gazed into the liver of a sacrificed sheep (see Chapter 7), the shared belief seems to be that the secrets of the future are already encoded within the living matter of the sacrificed being. Often a scribe would record the prognostication directly on the bone.

About 40,000 such inscriptions from China have survived as testimony to the popularity of the practice. They give us a good idea of the sorts of problems and questions confronted by the diviner and the manner in which they were worked out. When should we make a sacrifice? Will the harvest be a good one? Are the prospects sound for the king's hunting expedition? What will be the outcome of his forthcoming military raid? Will a peace treaty soon be drawn up? Once the practice of reading oracle bones became widespread (about 1400–1100 BC), manuals were written to give specific instructions to the specialist on how to read and interpret various patterns. Given our previous discussion of terms, it appears that a serious reader of bones would have functioned more as legitimate priest than bogus magician—at least to judge by our admittedly biased modern standards.

In Egypt the work of the state medium took place deep within the inner sanctum of the temple. Letters inscribed to the dead survive in papyri from the Middle Kingdom (*c.* 2040–1640 BC). "How do you fare now that the West [where the dead reside] is taking care of your desires?" reads one. And then a plea for action: "Remove the infirmity of my body ... [and] ... I will deposit offerings for you when the sun has risen ...."[5] Commoners were allowed to put questions to likenesses of their deified kings when they were paraded about on festival days. Shopping for a better oracle was not unusual: "Who has stolen garments from the storehouse?" asked the workman in charge. As the names of the villagers were announced one by one, the god flinched at the mention of a particular name. Protesting his innocence the accused vowed to consult the oracle of another temple. The papyrus recording the case indicates that the proceedings dragged on for weeks.[6]

Of all the oracles of the world none is more familiar than the oracle of Delphi. It was so popular by the 5th century BC that the line of future seekers could be quite long by midday. The medium was a woman, likely a member of the then ancient Mycenean cult of the earth goddess. Once asked a question and paid a fee she would enter a chamber at the back of the temple. There she would inhale smoke[7] that emanated from a vent in the earth and chew laurel leaves to induce a trance. She would then return and wail out a garbled response which would be interpreted and relayed back to the inquirer by one of her assistants in verse form, itself often charged with potential multiple interpretations. Some scholars believe that the image of a drug-induced hag ranting out irrational prognostications is a romantic invention and that what really went on was a consultation with interpretive prophets accompanied by a preparatory rite such as a goat sacrifice.[8] Less a magical device, the oracle simply may have been ancient Greece's analogue of the TV help-and-advice program, a sounding board onto which troubled people projected their concerns—a medium for motivating them to undertake specific action to resolve life's many problems.

Disappointed by the oracle's response to a question about what course of action to take in the wake of an anticipated invasion by the Persians (build a wall around Athens, replied the Delphic seer), the delegation acquired a distinctly different response from another oracle: "Divine Salamis, you will bring death to women's sons when the corn is scattered,"[9] reads part of a lengthy prophecy. Build a wooden wall of defense around the city or go on the offensive with a wall of wooden ships? Kick-started by their oracular consultations, the Greeks made their historic decision and, in one of their greatest maritime efforts, won the day at the battle of Salamis in 480 BC.

Today some members of Western society look down on crystal gazers and channelers, consultation via phone network and internet sites, but the status of the profession of divining from signs in nature in the ancient world was often quite respectable. In Assyrian culture, for example, the *dagil issuri* (bird watcher, or more accurately "bird operator") was regarded as an individual of high status. He would carefully study the movement of herons and eagles. Sometimes he would release doves or falcons kept especially for the purpose of divination and observe their flight patterns to extract his omens. Flocks of birds emerging from caves were a specialty. A text on Hittite ornithomancy dating from 1100 BC lists 27 species of bird. Some inquiries were said to have necessitated up to three days of bird watching. Lengthy manuals even offered advice on how one should or should not consult the *dagil*, for example, if it is by a bird that lives in a cave, then the sun god will be hostile, etc.

Ornithomancy was also an honored calling in India and Tibet. Crows and ravens were sacred in Tibet, thought to be messengers from the divine protectors. The *kakajariti*, a 19th-century text, offers a lengthy discussion of different types of crows and the sounds they emit. One must call out the question or affair as soon as the bird takes flight, then;

> When in the first watch a crow sounds its notes in the east,
> the wishes of men will be fulfilled.
> When its sounds its notes in the southeast,
> an enemy will approach.
> When in the south ... a friend will visit ...
> When in the southwest ... unexpected profit will accrue[10]

and so on around the directions of the compass.

The Roman augurs, or *auspicium*, inherited many of their skills via forerunners from the Middle East. On the occasion of the dedication of a temple, the best among them would be called upon to look for favorable signs. The 1st-century BC Roman writer Varro tells us precisely what the augur—literally one who divines by the observation of the course of birds in flight—did when he laid out the earthly templum. "When you look south for the seat of the gods, the east is on your left, the west is on your right. I believe it is for this reason that the augurs on the left are to be better than those on the right."[11] The concept of the lucky right, and unlucky or "sinister" left is indeed a Greek idea (curiously the Etruscans had it the other way round). In performing his rite, an augur first fixes a point on his left marked by a certain tree (which Varro goes on to describe). He marks out a corresponding point on his right. In the middle, directly in front of him, he determines the sacred boundaries of the templum "by sight and by meaning." Then he directs his glance over city and field beyond. He draws a line from east to west, then marks the limit in front of him as far as his eye can see.

Whether Maya or Tibetan, Greek or Roman, the quest for spiritual wisdom and the foretelling of specific events via the many "mancies" I've been describing are part of shared attempts to seek order in the real world—to search out the underlying harmony in which we all have faith so that we can undertake the most suitable course of action to fulfill our existence. Of all the natural media for conveying pristine order, perfection, and certainty, none exceeds the capacity of the sky. It is a simple fact that no other aspect of nature is more precisely predictable than the courses taken by the sun, moon, planets, and stars. You can depend on the sky. But acquiring foreknowledge via celestial phenomena is no mean task, for it demands not only a high degree of familiarity with the ways of the transcendent, but also great skill and persistence in perspicacious skywatching. As the inscription on the ancient Egyptian statue of a famous Egyptian astrologer (quoted in this chapter's epigraph) will attest, stellar divination was once an exalted profession.

Like all difficult and responsible callings, ancient stargazing had its stresses and anxieties. Wrote one frustrated Assyrian court astrologer unable to pin down a prediction because he missed a last disappearance of Venus:

> The lord of kings has spoken [to me] thus, Why hast thou not [observed?] the month and sent the lucky and unlucky. The prince of the kingdom has been neglected, has not been obeyed. May the lord of kings when his face is favorable lift up my head that I may make decisions and tell the king, my lord.[12]

The logic behind stellar divination is really quite straightforward. Through everyday experience a careful observer can easily become aware that the cycles of the sun and the moon are correlated with the seasons, the tides, the menstrual cycle. Why not extend celestial destiny to encompass tides as a force of influence in the affairs of people? If we watch the sky carefully enough, can we discover associations between the most precisely predictable occurrences on nature's stage, such as eclipses or heliacal risings of Venus, and the more vagarious—say a plague or the arrival of locusts? These seem to be the sorts of questions in the minds of antiquity's courtly time keepers, the sky specialists who composed and dictated the contents of astronomical tablets to their scribes. The language comprising the dialogue between mortal and transcendent consisted of offerings and incantations; the implements of communication were charm and amulet rather than compass, astrolabe, or sextant.

If, in the study of early astronomies, we overlook the deeper issue of motive and extract only the scientific principles of the exalted hand-me-down mechanics that lie at the foundation of modern astronomy (such as commensuration and the interval-place-interval scheme that we will discuss in Chapter 10), we risk losing an appreciation of how our ancestors comprehended the celestial sphere.

In most of the history of the world, astrology has been the generative force behind early astronomy. This was particularly true when, *c.* 800 BC, the Assyrian empire dominated Mesopotamia. Dynasties were tied ideologically to a pantheon consisting in large part of spirit deities whose actions, revealed through omens, influenced the course of people here on earth—especially that of the empire and its ruler. Omens are usually made manifest through the action of celestial bodies. They are the ends for which astronomy offered the means (Fig. 42).

The challenge of the astrologer-astronomers of old Babylon was to develop the skills whereby they could successfully read the omens and schedule the rites, usually to their tutelary god Marduk. Part of their skill set depended on good record-keeping and the invention of clever predictive formulae. And so they carefully scanned the skies above to follow their gods: "When the star of Marduk appears at the beginning of the year, [then] in that year corn will be prosperous."[13]

Whoever wrote the many "when ... then" phrases that make up the ancient corpus of omen-based astronomy must have been closely

42 *The crescent moon in Taurus (center), thought to be a position of great influence, with the stars that comprise the Pleiades to the left. Part of an astrological tablet from 4th-century* BC *Babylonia.*

connected with what was going on in the natural environment, one who witnessed nature's forces directly—earthquakes and floods, miscarriages and deformities at birth, eclipses, and rainbows. The ancient counterpart of today's astronomer did not conceive of such phenomena as detached events in a universe devoid of meaning (we will discuss just how that detachment came about in the Western world in Chapter 11). For the astronomer-astrologer, all things happened for a reason—to warn us and to convey a message to us, either good or bad, that would guide our future action. Some phenomena occurred with more predictable regularity than others. For the farmer a moonrise could tell what to anticipate in the forthcoming crop cycle; for the king an appearance of Jupiter might signal what an encounter with the people who lived to the east, those over whom he sought to extend his dominion, might portend. Avid skywatching, expressed via predictive, scientific astronomy based upon simple arithmetic, was passed on to the Greeks who refined and developed it and, as we shall also see in Chapter 11, gave us so much more.

Our word "horoscope" comes from the Greek *horoscopos,* meaning "one that observes the hour." It refers to the art of predicting general patterns that can be pre-programmed to suggest circumstances that might develop in your future life based on a careful inspection of the celestial bodies that appeared over the eastern horizon at the time of your birth. Unlike Babylonian astrology, which was restricted largely to affairs of state, in the Greek democratic system everyone had a right to a knowledge of the future. The system was quite complex, for not only did it extract information from the local positions and motions of the sun, moon, and planets with respect to the twelve houses (30° strips of the ecliptic beginning at the eastern horizon), but also similar divisions commencing at the vernal equinox or "first point of Aries" (one of the intersections of the ecliptic and the celestial equator). The latter gives us the familiar twelve signs of the zodiac discussed in the chapter on constellations. The first 30° sector of the

House system was the all-important House of the Ascendant. Other houses included that of Love and Marriage, Death, Honor, Friends, etc. Planets residing therein at the instant of one's birth were thought to exhibit the most potent outcome upon such matters throughout one's life.

The planets themselves possessed powers that alternated between good and evil, that power diminishing as one passed down the hierarchy of orbs. Planetary positions in the zodiacal signs carried additional meaning. To make matters more complicated, other properties of nature were tied to the signs. Gems, metals, herbs, parts of the body, individual organs, bodily fluids (humors), types of discharge—all were compartmentalized within a universal taxonomy that served as a kind of toolkit for divining. In the hands of an expertly trained astrologer, all entities that make up the ordered world could be brought to bear upon a prediction, whether it dealt with medicine and healing, the conduct of war, or how to handle grief over the loss of a loved one. Their system has been passed on to us in the watered-down popular form of the daily horoscope, and as anyone who has had their chart done (or done it themselves via the many software packages readily available) will observe, geometry is the dominant skill that underlies horoscopic astrology. The Greeks invented geometry and they regarded it as the most supreme form of logic. They were particularly aware of geometrical patterns that connected houses and zodiacal signs, e.g. they watched carefully which planets were in the Trines (sets of constellations 120 degrees apart, which form patterns of equilateral triangles on the horoscopic chart).

A fact often overlooked about modern astronomy is that many of the skills, especially the use of instrumentation employed in acquiring a precise knowledge of the sky, were not passed on directly to us from the Greeks via the Renaissance. Rather, they were developed and refined in Islam in between (especially AD 900–1400). A historian of the astronomy of Islam has remarked: "[One is left] amazed at the amount of theoretical, mathematical and astronomical knowledge that was produced in the service of religion, and the sheer beauty of the instruments rendered all that knowledge practicable."[14] Science and religion were certainly not in conflict in the Islamic world!

Following the downfall of the Classical world, many pious Muslims (including Mohammed himself) at first suppressed Greek astrology. Being both deterministic and monotheistic like the early Christians, Muslims tended to reject some of the fruits of early astronomy acquired from the Greeks, such as the casting of horoscopes, which came along with the scientific package of geometry and rudimentary astronomical instrumentation. Nonetheless, some of the greatest Islamic astronomers, among them Al-Biruni (AD 973–1048), practiced the occult art of astrology, which he seems to have regarded as more closely allied with scientific astronomy than we might believe today. In his *Elements of Astrology*, for example, he tells of the possibility of predicting meteorological events, such as floods and even earthquakes, along with the behavior of plants, animals—and people. But he admits that some predictions have origins that can never be known. When an astrologer passes this metaphysical boundary, Al-Biruni tells us, then he is on one side and the sorcerer is on the other. Operating very much in the medieval tradition of astrology, Al-Biruni seems to have believed in the tangible and scientifically detectable powers of astral influence, such as celestial rays that emanated from each sky object, causing us here below to vibrate sympathetically or antipathetically. Al-Biruni was a true stellar determinist.

Ancient Chinese astrologers were no less dedicated to their pursuits than their Middle Eastern counterparts. As we will recall (Chapter 2), the Chinese zodiac consisted variously of 28 or 36 houses arranged around the celestial equator, no doubt because the equator lies at right angles to the all-important fixed pole, which they revered as a symbol of their ruler's fixity and permanence. Their system stands in contrast to the Western tradition, which is based on the ecliptic. Traceable to sometime before the beginning of the 1st millennium BC, when it may have been imported from India, the early choice of 28 is obvious, especially if you want to distinguish the position of the moon among the stars on a nightly basis strictly for the purpose of timekeeping. By following the nightly crossings of the meridian of each of the constellation houses, Chinese astronomers were able to track the course of the months within the 365¼-day year.

Of what do these constellations consist? The earliest surviving star maps from 5th-century BC tombs give names of zoo- and anthropomorphic body parts, such as beak, mane, stomach, wing, heart, and gullet, while others are named after domestically related entities, such as harvester, house, well, ox, and winnower. Still others, like ghost and *triaster* (the three stars that make up the belt of Orion), seem more abstract. Despite any impressions we might acquire, these names turn out to be neither obscure nor irrelevant to prognostications about the real world. Thus, the mane (or yak-tail) relates to events pertaining to warriors, the net to hunters, the lasso to prisoners, the stomach to matters of the warehouse and granary. The (turtle) beak governs the harvesting of wild plants, while the ghost is capable of detecting cabals and plots against the emperor.

For every affair of state the starry winds of good and bad fortune blew across the sky. It was up to the astrologer to call the correct prediction. For example, one omen has it that when the stars of the well, which relate to all things quaffable, are faint, then wines will better diffuse their aromas; but when its brightest star is especially red, then dark red drinks (and food) may be poisoned. Another example concerns the entry of a bright planet into the constellation called Mao (our Pleiades) by the Chinese, the great sky sign of Tibetan warriors. An astrologer was thus said to have predicted the death of one of the most destructive alien invaders of the house of Tang.

Beyond the sun and moon lay the planets and their cycles. As in Greece, each member of the Chinese planetary pentad was identified with one of the five terrestrial elements: Mercury with water; Venus with metal (because it glowed "Grand White"); Mars, called the "Sparkling Deluder," with fire (for obvious reasons); Jupiter with wood; Saturn with Earth. When it came to timekeeping, if Venus was the fixation for ancient Maya astronomers of the New World, Jupiter was the clear choice in China. No one knows why, but I suspect it may have been the key to the lock on the commensurable long cycles that caught their eye: Jupiter undergoes grand conjunctions with Saturn in the same constellation every 60 years. This is an exact multiple (five) of the time it takes for the twelve earthly branches of *ti-chih* of the equatorial zodiac to make a full circle. (Popular Chinese culture still calls to mind the names of Chinese New Years: Rat, Ox, Tiger, Hare, etc., which are designated by the magical twelve.) But what may sound like number juggling to us has deep physical and moral meaning for those who fervently sought closed loops of time perfectly meted out by divine celestial bodies. To an astrologer it would have been quite natural to relate the twelve-year cycle of the "Year Star" to the twelve earthly branches, especially if he wished to acquire a prediction about the political turf in whose celestial station a planet resided.

Omens as far back as the Shang oracle bone texts mentioned earlier (about mid-2nd millennium BC) offer clues that Chinese astronomers had apprehended the special cyclic qualities of the planet Jupiter fairly early:

> The divination on day *hsin mao* was performed by *Chi*. The king is to make a sacrifice to Jupiter. Will it not rain?
>
> The divination day *chi wei* was performed by Hsing. The king is to make a sacrifice to Jupiter. Will the offering of two oxen be sufficient to stop the disaster?"[15]

Hundreds of years later the Chin Shu family history gave an exhaustingly long list of the exact positions of Jupiter in various houses together with the sub-prefixtures affected by each particular association.

Change was the focus of Chinese planetary observations. Every subtle cosmic alteration had a name and a meaning. Each bend, kink, and turn, every slowdown or rapid motion the wanderers made along the sinuous celestial skyway was duly noted, for then their essences descended upon us. Particularly fascinating were the close conjunctions, especially the ones that took place during retrograde segments of their tracks along the zodiac. To give a few examples, it is said that when the five gather in the east the signs are for the kingdom in the center, and when they assemble in the west the omens will concern foreign countries that threaten the center. When two planets lie in the same constellation they are said to be in combat. The closer together, the greater the magnitude of the calamity. If Jupiter and Venus enter into combat, the confusion was likely to be of a civic nature. Venus moving to the south and Jupiter to the north in close passage had its own particular designation, while the opposite passage had another name. One meant a good harvest, the other bad. Mars-Venus conjunctions held still other, quite separate consequences. And when three planets arrest their motion all in the same place—then there will come a war with a heavy death toll.

Our discussion of the Maya codices in the chapter on "The Timekeeper's Sky" (Chapter 10) will reveal an intensity of focus on the part of their astronomers that rivaled that of the Classical world and the Middle East. For example, the lines of the Venus table that we will discuss in that chapter betray the motive behind the quest for a precise astronomy. A dozen glyph blocks, arranged over one of the effigies of the Venus deity in the Venus Table in the Dresden Codex, tell what happens following his first morning appearance:

Tied to the east,
is God L, great star [Venus]
Kawil is speared
woe to the twenty, woe to people
is the divination, two-blue yellow
woe to the Maize God, woe to the food.[16]

These statements regarding the consequences of the return of a celestial god after his brief visit to the underworld exhibit the same "when … then" formula we found in Babylonian texts, like those about Marduk, or Hittite and Etruscan omens derived via bird augury. Make no mistake: in all these cultures the quest for astronomical precision is driven by religious beliefs—a fact that is very difficult for the student of contemporary science (or religion) to understand.

Maya omens sound a ring of fatalism, almost a kind of resignation to the forces of nature. The ills that befall the world all seem to be aligned with particular sets of deities. One needed to know what these gods were up to so that the proper course of action—usually some sort of sacrifice honoring a reciprocal contract—might be performed. Only human action can keep the universe in equilibrium. Sadly, we know next to nothing about how ancient Maya daykeepers, or *Ah K'in* (after the root work *k'in* which means sun and day as well as time), actually operated, for no astronomer's notebooks survived the great document and book burnings of zealous Spanish priests; but some vivid descriptions of calendrical divination are extant (Plate 5).

The contemporary Maya daykeeper appears to be in tune with the gifted diviner of half a millennium ago with whom he shares his craft. Anthropologists tend to identify the diviner as a shaman because he/she works individually, and is recruited in typical shamanic fashion—by being born on a special day or having particular visions or special dreams. The gifted one is said to receive "lightning," "a kind of soul … that enables him or her to acquire messages from the external world"[17] through bodily sensations. But unlike his predecessor from the age of the Classical Maya, today's diviner carries no codex. He/she sits at a rock table with special apparatus aligned to the four directions. These include lighted candles, incense bowls, and lines of little piles of rock crystal and groupings of seeds. The crystals and seeds are manipulated in order to divine answers to questions based on a dialogue with the days of the 260-day calendar. For example, if the day of the divination is 1 Quej, the shaman might address the piles directly, asking literally to "borrow the day"; thus "Come here, 1 Quej, you are being spoken to."[18] If the client is ill the shaman may ask the day the illness began to stand in for 1 Quej, so that he/she may conduct a dialogue with it. A mysterious (to us) mathematical reconstruction of seed and crystal redistribution yields up the answer to the question: "What will be the fate of my illness?" All these divinatory rites are backed up by a shared complex astronomy that exhibits a close understanding of the cycles of the sun, moon, and planets.[19]

Contemporary astrological shamanism persists in many communities removed from modern globalization. Cheyenne shamans of the U.S. Great Plains are thought to have special access to the multiple regions of the layered universe (Fig. 43). The most powerful region of all, Blue Sky Space, is closely watched, for the constellations there give subtle signals. Ursa Major or "Kheglun" is the moose, and Bootes, or "Mangi," the half-animal (bear), half-human who pursues him. Mangi is said to be the first ancestor from whom the shaman acquired via dreams his transformative powers that enable him to fly among the stars.

In preparation for the sacred hunt, the Cheyenne shaman elicits the power of the "Blue Star" (Rigel). An elaborate ritual called the Massaum, or crazy buffalo dance, is recounted in detail by an early 20th-century traveler.[20] They say that into the bowels of the earth descended a man and a woman charged with the task of saving their people from starvation. There from the Great Spirit they received the first Massaum ceremony as a gift for providing game animals to feed

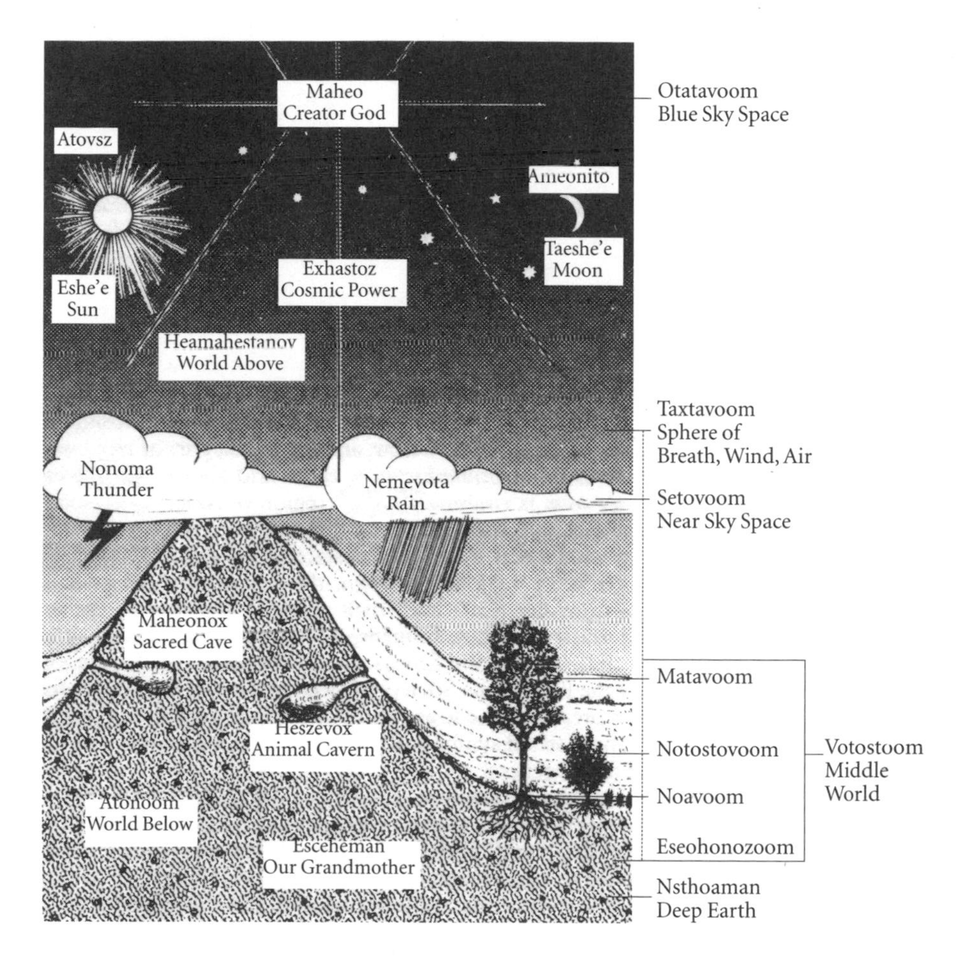

43 *The multi-layered universe of the* Tsistsistas *(Cheyenne) of the U.S. Great Plains is a complex hierarchy of layers to which sacred places (caves, mountains, thunder, rain, etc.) are assigned. At the top of the hierarchy lies Blue Sky Space, the destination of the shaman who takes flight. It is said to be the root of cosmic power.*

the people. Re-enacted to commemorate the event, the dance follows after an elaborate circular seating plan is set up in the camping lodge in accordance with the four directions. Five special shamans leave the circle. They stand facing southeast waiting for the signal to emerge from the sky: "Finally, the blue beacon of Rigel blazed above the dark horizon. They stood and watched. For a few minutes the star shone intensely, then dimmed and quietly disappeared."[21] Masked animal

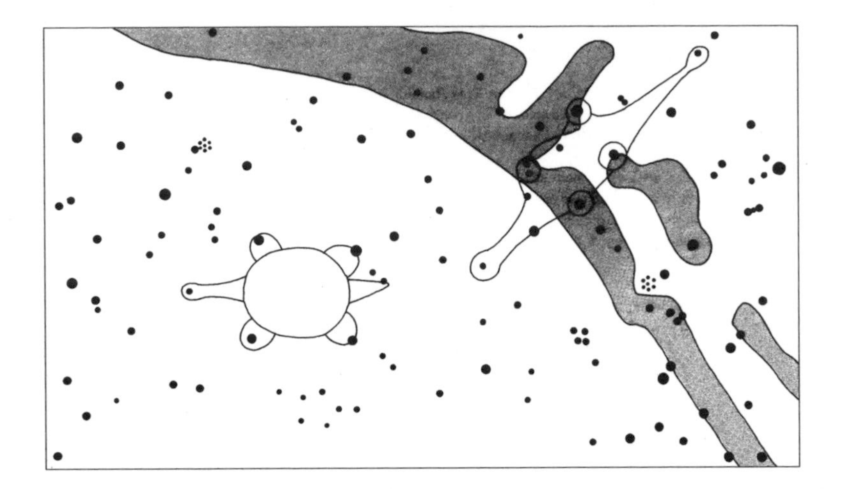

44 *Lakota constellations of the turtle and the salamander (they are made up of stars of our constellations of Pegasus and Cygnus, respectively). Once the umbilicus of a newborn is cut it is thought to be symbolically reconnected to one of these star groups; for a boy, the salamander, for its agility and adaptability; for a girl, the turtle, for steadfastness and long life.*

impersonators—buffalo, elk, deer, antelope, otter, wolf, crane, coyote, badger, cougar, fox, eagle, hawk, raven, magpie—all emerged from their dens and danced in a huge ring in clockwise fashion. The head shaman appeared in their midst and announced "I am the star of dawn."[22] Thus he acquired from the power of the Blue Star of the highest Blue Sky Space, the status of master spirit of the animals. Let the hunt begin!

Lakota midwives, also called to their work through dreams, are regarded as shamans who possess the power to access the spirits of the stars. Once labor begins a midwife is summoned. She brings special plants to curtail hemorrhaging or to speed up the arrival of the placenta. Then she prays to "birth woman" (or Blue Woman), who comes out of a hole in the sky located in the bowl of the Big Dipper. Birth woman guides the baby's spirit into this world and eases pain during the delivery process. She reincarnates spirits in the material world. "And then, after death, she aids those same spirits in their passage out of the material world back through the opening in the dipper into the spirit world, their place of origin."[23] Once the baby is born prayers are offered to the turtle or salamander spirits (constellations) (Fig. 44), asking them to bestow their power on the child.

A final example of the art and craft of contemporary celestial divining comes from the Aborigines of Australia, where the power of the stars is often directed toward healing and medicine. The Waradjeri shaman-healer takes a skyward journey that begins with a song to the clouds, that they might come close to the ground. Lying in a coiled position he sings himself up to and past them, like a spider moving along a vertically spun cord. This vivid description of the shaman's vision survives:

> He could see the darkness of the night sky, and all the stars, which were the Ancestral Beings who had in the past climbed up here; being so close to them he could see their human forms, whereas from the earth they appeared merely as points of light of varying brilliance ... To pass into *Palima* (the place where the water-bags were kept), the doctor had to go through a fissure, through which the Ancestral Beings had passed when they left the earth. This fissure or cleft ... (had) two walls ... continually moving around ... On one side ... sat the Old Moon Man. He had a long beard which reached to his waist, while his penis was so long that he had to bring it up and wind it around his waist several times wearing it thus as a waist band. On the other side was the Sun Woman; she had protruding breasts, and sat in such a way that her large distended "labia majora" revealed an extraordinary elongated clitoris, which covered the fire made by the sun and the daylight ... His (the healer's) actual journey was said to have taken no more than a few seconds.[24]

Endowed, like most shamans, with the power of direct contact with the ancestor gods, the shaman returns to earth and conducts the appropriate healing rite over the victim. Invariably accompanied by a song of exorcism, the procedure might include massaging his own sweat into the body of the afflicted one, or tying a cord to the sick person's foot while unraveling the other end into a stream in order to offer an exit path to the invading spirit.

If some of the astrologer-shaman's mystical revelations I have recounted in this chapter cause the rational mind to bristle, consider the sensory revelations recounted by modern cosmologists who address their inquiries to the galaxies that populate our universe. Astronomer Sir Fred Hoyle, for example, tells the story of his struggle with a mathematical problem that had so defeated him that he decided to quit it and take a vacation. As he rode the train to his destination the problem suddenly intruded itself upon his mind. Devoid of any writing implements, he fiddled with it. Suddenly, he writes, "my awareness of the mathematics clarified, not a little, not even a lot, but as if a huge brilliant light had suddenly been switched on. How long did it take to become totally convinced that the problem was solved? Less than five seconds,"[25] answered Hoyle.

Einstein spoke of a "cosmic religious feeling" that lay behind some of his eureka moments concerning the harmony of nature; still others have spoken of "an overpowering force," a power that is "volcanic, pent up, ready to be unleashed," or "a remarkable feeling of intensity that seems to flood the whole world around us with meaning ... We sense that all boundaries between ourselves and the outer world vanish, for what we are experiencing lies beyond all categories and all attempts to be captured in logical thought."[26] Such descriptions, collected from among a number of well-established scientists, may lack the colorful imagery reported by Maya or Cheyenne shamans. Nonetheless they convey that same spontaneous sense of transcendent holism directly apprehended—a moment of inspiration arrived at not via any process of rational thought, but rather all in a flash, immediately. Perhaps there is a way to the future that is shared by both shaman and scientist.

CHAPTER 10

The Timekeeper's Sky

Whatever next? Scientists at the NPL are also working on a new form of time standards, known as an 'ion trap,' which is tentatively predicted to be correct to 'one part in ten to the power of eighteen.' This surely has to be an absolute time source as it would be accurate to one second in ten billion years—or accurate to one second during a period that is believed to be the lifetime of the universe itself.[1]

A historian of timekeeping

We have a tendency to interpret the past by the standards and norms of the present. When it comes to understanding the history of science, our dependence on technology often leads to unwarranted assumptions about the scientific achievements of our forebears. This is especially true when it comes to the development of sophisticated timekeeping systems in antiquity. How did they achieve such accuracy without telescopes to pinpoint the exact positions of stars? How did they keep track of time prior to the invention of wheeled gears and mechanical clocks? How did they perform the mathematical calculations required for accurate timing without mechanical calculating devices or computers?

Another question frequently raised about precise systems for keeping track of time in antiquity is: why create them in the first place? Did ancient people really need to be so precise? What purpose did precise knowledge serve? In this chapter we'll entertain these questions by looking at contrasting stories reflected in two of the most ancient precise timekeeping systems—systems that developed in total isolation from one another. We'll focus on who the timekeepers were, what drove them, and the role that the sky played in their culture's mandate for accuracy.

My first tale extols Western civilization's penchant for precision, and deals with the development of our own seasonal calendar. We are

Tools of the Timekeeper

Make no mistake: there was technology. Invented by the Greeks, the astrolabe (meaning "star-taker") was one of a number of astronomical measuring instruments developed in Islam that lay at the source of the precise data that appeared in collections of astronomical tables all across the southern Mediterranean world from the 10th to the 15th centuries, before the Mongol advance from the northeast left the smoldering ashes of Muslim science for Renaissance European scholars to pick over.

Combining the sighting properties of telescope and the figuring capacity of computer, the astrolabe was basically a mechanism used to tell time. The viewer looked by day or by night through a pair of sighting holes at opposite ends of a rod mounted on a circular plate that looks like an overgrown pocketwatch, hand-held by a long chain. (Actual dimensions ranged from a few inches up to a foot-and-a-half—see accompanying figure). The face of the watch was outfitted with a series of star-map plates that could be removed and substituted like compact disks, one for each appropriate latitude to which the user might journey. Each of these disks consisted of a flat stereographic projection of the sky onto the celestial equator (the extension of the earth's geographic equator onto the sky). The central hole marked the fixed position on the celestial sphere, approximated by the Pole Star. On the top of this, another plate gave basic coordinates in horizon, equatorial, and ecliptic systems (the smaller circle in the figure is the ecliptic or the plane of the zodiac). Positions could be read off the rotatable ruler once the appropriate object was sighted through the peep holes. Another circle on the flip side of the star clock the benefactors of the old Roman calendar, which, though it was first written down about 300 BC, probably dates back to the 6th-century BC founding of the Republic. It began quite simply as an agrarian routine organized by those closest to local tribal chiefs, consisting of a simple listing of festivals and *fas* or "court days" for conducting business. It was not until it passed into the hands of urbanized society and became subject to the demands of many interests—trader, merchant,

served to fix the hour of the night, the day of the month and the position of the object sighted on the zodiac; the latter function offers a tentative clue to one of the more mundane uses of such an instrument—astrological prediction.

Of all the parts of an astrolabe, the pointers are the most curious, often appearing as tongues or talons of mythical creatures. There also exist astrolabes with birds, bulls, horses, whales, and bears gesturing and posturing on their clock faces. In one example warriors dance about as the tips of their swords do the pointing, while another, shown in the figure, uses dog heads, the protruding tips of their panting tongues excitedly showing the way of the stars (appropriately enough, Sirius, the Dog Star, is a major feature on this particular instrument). The source of New World (e.g. Maya) tools for astronomical observation is more problematic. Nothing that looks like an astronomical instrument has ever been excavated or reported upon in print. But pictures from the codices that show the head of a man perched in a temple doorway looking through a pair of crossed sticks are suggestive.

This 15th-century astrolabe, a device for telling time by the stars, employs the claws and tongues of panting dogs as indicators of star positions.

artisan, bureaucrat—that the calendar was placed in the hands of specialist scribes, whence it became formalized, complex, and subject to irrational vagaries. It also began to be lengthened.

The Roman calendar originated with the two most obvious cyclical astronomical referents: the sun, which gave us the day, and the moon, whose approximate 30-day cycle of phases yielded the month. In addition an eight-day market week got meshed in with nature's cyclic pair to produce a 120-day year, or *annus*, that consisted of four months bearing the names of deities: Martius, Aprilis, Maius, and Junius.[2] Later, six months were added (Quintilis, Sextilis, September, October, November, December). Their names, quite unlike those of the months that preceded it, signified their numerical order: fifth, sixth, etc., though the first two later were changed to July and August after Julius Caesar and the emperor Augustus. This form of the year, composed of ten lunations or 300 days, was fashioned to correspond roughly to the gestation period of both humans and cattle, whose life cycle occupied a fair share of agrarian Rome's attention.[3] Because it would be more orderly to keep the weekday names in step with the month, i.e. to begin each new year on the same day name, four additional days were added later (one each to the months of March, May, July, and October). This brought the number of days in the year to a total of 304, a number divisible conveniently by the eight-day week without remainder.

Strange as it may seem to us, the remainder of the seasonal cycle was simply not tallied, falling as it did during the portion of the year when the fields lay fallow; thus the idea of the year being equated with the full seasonal or solar cycle had, as yet, no place in history.

To this point it seems clear that one motive that drove the keepers of our early calendar was the quest for commensurate periods, or time cycles that interlocked together harmoniously, that is, in the ratio of small whole numbers. Perhaps to convey a sense of better control over time, a second motive was the desire to incorporate natural periods of longer duration into the date-reckoning scheme. These lengthier event sequences could also serve as checks against one another. Indeed, we would feel far more secure if we could depend on a kind of order discovered in the movement of both the moon and the sun cycles rather than only one of them. The first attempt to coordinate the lunar month and the solar year occurred around 750 BC when Numa Pompilius, the second king of Rome, added 50 more days to the 304-day calendar.

In this new and extended calendar, eight days were subtracted from among the other months, this sum being added to 50 to make a 58-day period, which then was divided in half to form two new 29-day months: January and February. These were tacked on to the end of the month list to make a twelve-month year, the closest approximation to a solar-seasonal year of 365 days that would accommodate a whole number of lunar synodic months. By 150 BC an additional day was added (to the month of January), perhaps to keep the total number of days in the year odd (which was considered lucky); and so the Roman year grew to 355 days in length. But this number would not wholly accommodate the traditional eight-day market week without the occasional insertion of days into the calendar, a process known as "intercalation."[4]

It took nearly 600 years from the time of the inception of the calendar before Roman chronologists embraced a challenge that would occupy them, if not with equal intensity at all times, through the zenith of their empire: how to devise a manageable, lasting method for inserting days into the calendar in such a way that the canon of time they had created would keep in step precisely with "real" time marked out on the sky by the moon and sun cycles.

Basically there are more than twelve and fewer than thirteen lunar phase cycles in a year—12.3683 to be exact. The easiest way to create a seasonal calendar that brings the two together might be to reckon two years of 365 days followed by one year 366 days in length. Instead the

Romans chose to add a thirteenth month, either 22 or 23 days long, every *other* year. But this posed political problems for the pontiffs, who were responsible for the administration of the cults of the state ordered by the calendric observances; for example, some prelates allowed the lengthier years to be those during which their friends were in office. It also posed calendrical problems. Suppose, for example, that we have four sequential years 355 + 22, 355, 355 + 23, and 355 days in length, making a four-year total of 1,465 days. Now, four tropical (seasonal) years, measured by the time it takes the sun to make successive passages across the vernal equinox, equal 4 × 365.2422, which is just minutes short of 1,461 days. Thus the festal days would gradually began to slide through the seasons. By the end of the 2nd century BC skywatchers begin to notice the difference via progressive changes in the seasonal disappearance dates of the zodiacal constellations in the west after sunset.

The problem was not confronted with any rigor until Rome was pressed with the need to set reliable time standards over the ever-expanding territory of their empire. Sosigenes, chief advisor on calendric matters to Julius Caesar, recognized that the shift of the sun's position in the heavens relative to the dates being tallied in the Roman calendar had built up to an astonishing 67 days. He suggested a set of remedial measures which since have collectively come to be known as the Julian calendar reform: First, he made an adjustment in the year count to restore the date of passage of the sun by the vernal equinox to the correct position in the calendar, then 25 Martius. This necessitated the intercalation of three whole months in 46 BC, a year which came to be known as the *Annus Confusionus* or "The Year of Confusion"—a 445-day year! Second, Sosigenes abandoned the moon as one of the basic cycles of nature by introducing twelve alternating 30- and 31-day months that added up to a year of 365 days. The 31-day months were January, March, May, July, September, November. (Later, the emperor Augustus, not to be outshone by his predecessor, removed a day from Februarius and added it to his own month.) Sosigenes' action made it desirable, if only for the purpose of balanced alternation, to convert October and December to 31-day months and September and November to 30-day months, thus making the calendar quite recognizable to the one they passed down to us.

So ended the habit of month intercalation and, for the first time, the only natural period that would serve as the fundamental time unit in the calendar was that of the sun. Finally, Sosigenes instituted a third, and most important measure. To assure that no further slippage between real and canonic time would occur, every fourth year he added one day to the year—at the end of Februarius—so that, averaged over a four-year period, the length of the canonic year would become 365.25 days, a very close fit to the real-time solar period.[5] But even with the day intercalation program of the Julian reform in place, solar-based events still fell behind the tally kept by the chronologists by approximately one day in a little over a century—though such a trivial amount was hardly worth worrying about given the temporal quagmire out of which the Roman timekeepers had just extricated themselves.

The battle with nature over precise timekeeping was not resumed for a few centuries, but this time it was fought on the fields of the ecclesiasts. The issue had to do with scheduling the celebration of the Easter holiday in the new Holy Roman empire. In the 4th century AD, with the emphasis of unity in the Roman empire now being placed upon the shoulders of the Christian Church, pagan cults and rituals, along with the old Roman calendar, were abolished. In their place, a calendar based on ecclesiastical reckonings and the commemoration of saints was instituted. One goal of the Council of Nicaea (today Iznik, in Turkey), held in AD 325, was the setting of a single date for the celebration of Easter by Christians in both the Eastern and Western Holy Roman empires. Traditionally the date had been fixed in the Hebrew calendar as 14 Nisan, or the fourteenth day of the first lunar month, which began with the first sighting of the waxing crescent moon nearest the spring equinox. But the Hebrew calendar, being based on the lunar cycle, did not observe the equinox with any special precision.

The Christians were compelled to celebrate the anniversary of the Paschal date close to the time of year when the event was documented to have occurred—as well as on a Sunday. Furthermore church law dictated that it must not fall on the Hebraic Passover. The "computists," or specialists in charge of calendric computations, believed a conflict could be avoided if they selected the first Sunday after the full moon that followed the vernal equinox. There were symbolic reasons as well: the light of Christ derived at the Resurrection is symbolized by the first period of full continuous light—twelve hours of daylight followed by a full moon visible all night. Contrast this with Christmas, which takes place at the maximum period of darkness in the seasonal cycle, symbolic of Christ having been born into a world of darkness and leading his believers to light.

Now, since it had already been recognized that the equinox was falling earlier and earlier in the calendar year (it had shifted backward three days in the four centuries that had elapsed since the Julian reform), the religious problem of determining the Paschal date became tied to the astronomical problem of fixing the vernal equinox.

Given the set of rules for determining the movable feast, the computists once again triumphed by reconciling three periods that do not commensurate very readily: the week (now seven days in length), the lunar synodic month, and the solar year. It turns out that Easter could fall any time between 22 March and 25 April; but it would take over five million years before dates of Paschal observances would recur in the same order. Realizing that it was impossible to devise an analytical formula that could be used to set dates in the future, the computists were forced to draw up elaborate tables based on averaged full-moon intervals. These tables of "epacts" listed the age of the moon on 1 January, from which the Easter date could easily be computed for that year. But these artificial tables gave only approximate information about the phase of the moon on New Year's Day (the error could amount to two or three days).[6]

At this point readers may question the need for such precise time reckoning. Indeed, one of the calendrical debates that followed after the Middle Ages centered around the extent to which religious festivals needed to be calculated with astronomical accuracy. After all, as astronomer Johannes Kepler put it, "Easter is a feast, not a planet."[7]

Time marches and error accumulates. By the 16th century the recession of the true equinox date, compared to the canonic count, had ballooned to eleven days. Flowers seemed to bloom too early and Easter Sundays became warmer and warmer. In 1582 Pope Gregory XIII appointed a commission to deal anew with the calendar reform issue. As was the case 1,500 years before, action needed to be taken to insure that the future festival date would fall on the proper date in the seasonal year: first, the equinox needed to be restored to its proper place in the time count; and second, the commission needed to devise a mechanism to hold it fixed (Plate 9).

After much debate about whether the lost time might be made up in small parcels over a long interval, the problem was solved in a single bold stroke (as in Caesar's time) simply by chopping ten days out of the calendar. Thus the equinox was moved from the 10 March date on which it occurred in 1582, to 21 March. This was accomplished by a papal decree declaring that Wednesday, 4 October be followed by Thursday, 15 October. We can only imagine the impact such a change would have upon the business and legal world should it occur today. How would salary periods be altered? When would rents fall due? How would quarter-year fiscal periods be reckoned? It is very easy to draw up a long list of problems, many of which were actually encountered four centuries ago.

The second step in the Gregorian reform consisted of changing the leap-year rule by decreeing that, among century years, only those divisible by 400 shall be leap years. Thus, while AD 1600 and 2000 would be leap years according to the new system, AD 1700, 1800, and 1900 would not. Such a scheme has far-reaching consequences, for it drastically reduces the shortfall inherent in the Julian leap-year system by cutting the length of the calendar year, averaged over long periods of time, below 365.25 days to 365.2425 days—only 26 seconds longer than the true year! So clever was the new rule that the canonic scheme would now pull ahead of the seasons by just one day in 3,300 years.

The man behind the daring plan was Jesuit mathematician and computist Christopher Clavius (1538–1612). Dubbed by one science historian the single most influential teacher of the Renaissance, his name appears on Pope Gregory's tomb, and one of the largest craters on the moon was named in his honor. Given the rudimentary state of computational mathematics at this time, we can appreciate such tributes. Calculating the time of the vernal equinox and the amount one needed to shift was a monumental task. It took Clavius 800 pages to explain it all!

The Gregorian reform was adopted by all Catholic countries immediately; but as might be expected, it was resisted rigidly by Protestant-dominated countries and at first given little attention by non-Western cultures. Great Britain and its colonies did not adopt the new calendar until 1752, by which time they needed to drop eleven days from the count. (Indeed, in early life George Washington celebrated his own birthday on 11 February instead of on 22 February.) The calendar was not accepted by Russia until the Bolshevik Revolution in 1917, by which time even more days needed to be eliminated. It seems that the more complex the bureaucratic interest, the greater the concern to build up and interlock larger and larger time cycles together. Minor reforms that further extend the accuracy of the time base have taken place since Pope Gregory's day.[8]

Current attempts to capture nature's cycles of time have begun to penetrate the rough edges of the cycles themselves. Any new rules that we could devise to further improve upon our accuracy would prove futile as we know now that the length of the tropical year itself changes, due to gravitational perturbations upon the earth by other bodies in the solar system. The rate of change is comparable to the difference between the true and canonical year.

To meet our modern needs for precise timekeeping, we still make minuscule corrections to guarantee that civil time, today defined by motions of atoms ("atomic clocks") rather than by motions in the heavens, will run smoothly. Because the rate of rotation of the earth, which defines the day, slows down by an average of .014 seconds per day every century, to keep non-uniform rotation time in phase with uniform atomic time, we add an occasional "leap second" to our clocks at the end of the sixth month.

And so, our modern calendar was not invented at a single stroke of genius. Instead it grew—evolving from simplicity to complexity, from approximation to precision. And while politics, economics, and religion color the narrative of the story of our encounter with time, at the base of this curious confrontation with nature lies the eternal quest for harmony, or as we term it in the realm of the timekeeper, commensuration.

I have gone to great lengths to tell the detailed story of the origins of our Western calendar for two reasons. First, to impress upon my readers the extraordinary precision that lies at the foundation of the canons of Western timekeeping, and second to convey how much non-astronomical detail may be unrecoverable from inquiries into non-Western timekeeping, wherein we lack the extensive written historical record that is available pertaining to our own timekeeping system. But Western computists were not alone in honing a penchant for precision. Their quest for commensuration was shared by the

Maya scribes, New World computists who pursued the vagaries of time on the other side of the Atlantic. The tale of how their colorful calendar developed, also worth telling in as much detail as the evidence permits, reveals some remarkable likenesses about how people reckon time.

An expert once characterized the Venus Table in the Dresden Codex (Plate 6) as a "subtle and mechanically beautiful product of the Maya mentality."[9] Painted with mineral dyes on the lime-coated bark of the ficus tree (native to Yucatan in Mexico), the document dates from about the 14th century. Its arrival in the library of Dresden was preceded by a long journey from the New World that began in the days following the Spanish Conquest, when, in the aftermath of the destruction of heretical material by the Viceroy, it was likely seized as booty. Certain pages of the Codex seem to represent an ingenious attempt by the Maya to create a perpetual Venus calendar, designed to predict the arrival of that planet at various stations of its cycle. The table is accompanied by a set of minuscule corrections and it is accurate to one day in 500 years!

As one investigator has suggested, the Venus Table, like all Maya ephemerides (an *ephemeris* (singular) is a table that gives the places of celestial bodies at regular intervals), was a "field manual for the secular priest containing applications of Maya science to the practice of the priestly profession."[10] The work of these priests, like that of the Roman pontiffs and the Christian computists who followed them, consisted of precisely setting the dates of celebrations and festivals and indicating the times of good or bad fortune for undertaking a particular action.

Just how does the table work? It lists the times of arrival of Venus at each of four stations: the first appearance as morning star, the last appearance as morning star, the first appearance as evening star, and finally the last appearance as evening star. As we shall see, the structure of the table implies that the first event held special importance.

These numbers, as well as the accompanying pictures of various forms of the Venus deity, and statements in hieroglyphic writing pertaining to matters of agrarian interest, suggest that, as in the problem of pinning down the Paschal date in the Old World, the considerations were not confined purely to astronomical matters. For example, at the bottom of each of five pages that resemble the one on the right in Plate 6, the intervals between each of the stations are tabulated (236, 90, 250, and 8 days, respectively) in base-20, as was the custom in all Maya mathematics (see Box: *A Maya Math Primer*, p. 206).

Now, except for the shortest one, these canonic intervals lie far from the observed values. They were probably set in place to guarantee, if not the actual observation date in the 365-day year of the arrival of the planet as its station, the celebration of that event on particular name days in a ritual cycle known as the *tzolkin*, or "count of days." The *tzolkin* of 260 days consisted of the matching of thirteen consecutive numbers with twenty consecutive named days (see Box: 260*: The Maya Number of Time*, p. 208). Finally, the other three intervals are close enough to integral and half-integral lunar synodic months to hint that they may be a vestige of an earlier, lunar-based calendar.

The challenge confronting Maya calendar computists is similar to the one that faced their counterparts in the Old World who developed our calendar: to create a canon that would keep the tabular reckoning of Venus (rather than solar) time in step with real time.

Now, the Maya reckoned the Venus synodic period as 584 days (the sum of the four intervals mentioned above), and the reading of the table indicates that they sought to keep that canonic Venus count in step with a great cycle of 37,960 days that commensurated all three relevant periods: the *tzolkin*, the Venus cycle, and the seasonal year.[11] (Regarding seasonal time, it happens that the synodic period of Venus (583.92 days) stands in relation to the year almost exactly in the ratio of 5 to 8.)[12] The great cycle represents the time elapsed by one full passage through the Venus table.[13]

A Maya Math Primer

Our base-10 system surely must have evolved from our pre-literate ancestors' habit of counting on the fingers. But a full body count in the Mesoamerican tropics was composed of the fingers and toes, with subunits consisting of hand-and-footfulls of whatever article was being counted. This is why we see dots (= ones) and bars (= fives) as symbols of quantity in the Maya codices instead of the more familiar base-10 Arabic numerals. But like our very own Arabics, borrowed from 6th to 10th-century Islam, the Maya symbols likely originated as hand gestures, ones being the abstractions of tips of the pointing fingers and fives the hands extended horizontally with the fingers brought together (which resembles a bar).

Base-20 further means that every higher order in the system is cast in powers of twenty rather than 10. Thus, in a vigesimal system the number 8.2.0, written in Maya, would stand for zero ones plus two twenties plus eight twenty-times-twenties. (Expressed in our decimally based system, this would add up to 3,240.) Note that the zero in the lowest numerical order is positioned at the bottom in this vertically arranged notation. Using a zero gave the Maya an enormous computational advantage over their European counterparts, at least prior to its introduction there in the Middle Ages via Islam.

One essential difference in Maya, just as in Babylonian notation, is that when time, as opposed to articles or things, is counted, the notation scheme is altered; thus, the third place becomes 360, or 18 × 20, rather than 400. This would have made the counting system more user-friendly when tallying units of days that make up seasonal years. Thus, if 8.2.0 were intended to represent counted days rather than, say, cacao beans, it would translate, in our currency, to 2,920. You are now equipped to read the numbers in the Maya document shown in Plate 6.

In order to make the table recyclable, Maya computists calculated a set of base dates consisting of named days in both the 365- and 260-day counts (together called the Calendar Round) on which a heliacal rising of Venus was predicted to occur. Ritual restriction dictated that the name day in the 260-day count that opened every great cycle must be 1 Ahau, the day name of the Lord of the Morning Star.[14] (Think of the 1 Ahau restriction in the same vein as the Christian restriction to Sunday for Easter.) Same stage, different actors: the problem really is exactly the same as the Paschal date-reaching problem in the Christian calendar, except that instead of dealing with the moon and the sun, here we deal with Venus and the sun.

Four lines of days in the 365-day year are written in the Dresden Venus Table. The first of these Calendar Round dates is the primary base—the one that first launches the user into the table. The others appear to be the result of some sort of shifting mechanism (the adding or dropping of days in the count, just as was done in our calendar) to correct the table to the true motion of Venus for future passages through the table. Such a shift would have been necessitated by the tiny, ever-accumulating observed difference between the canonic count of 584 days and the true Venus period of 583.92 days. Thus, like the equinox sun in the Roman calendar, the planet Venus would arrive at its station about five days ahead of the tabulated date every great cycle if the table were left uncorrected.

But the ingenious Maya timekeepers discovered a convenient way to initiate the correction process without violating the restriction that Venus must return to the sky on the sacred day 1 Ahau. They used a multiplication table (located on the right side of the lefthand page in Plate 6 which lies adjacent to the five-page table) of whole and near-whole multiples of the Venus synodic period to truncate the count in the great cycle some time before the 65th and final tabulated Venus revolution (see note 14). As in the "Do not pass go" command in Monopoly ©, users were then instructed to drop a specified (small)

260: *The Maya Number of Time*

The number 260 is as seminal to Maya time thought as is our gravitational constant to the laws of physics. Why 260 took hold as the foundation of Maya calendrics remains a mystery (it acquires prominence nowhere else in the world). This number is the result of multiplying the number of layers in Maya heaven (thirteen) by the number of fingers and toes, the basis for the Maya vigesimal system (see the previous Box). But 260 is also both an approximation to the period of gestation of a human female (average 253 days), and to the average length of the agricultural season in Yucatan. Astronomically, 260 days beats in the ratio of two-to-three with the long-term average interval between eclipses, the so-called eclipse year of 173½ days[15] and it is nearly equal to the actual average length of time Venus spends as morning or evening star (263 days). Such natural coincidences may have made 260 the number of choice for working out certain pairings of day names and numbers that were assigned to eclipses, the same combinations that would surface in future cycles; for example, eclipses will tend to fall only on or close to certain named days in the sacred calendar. All of this fits rather well with the Maya belief in lucky and unlucky days—that certain days, by their very designation, already carried with them particular good or evil influences from the past.

For all these reasons combined, the 260-day period emerged early on (*c.* 600 BC) in the development of Maya culture as nature's time cycle *par excellence.* It encapsulated the powers of all the gods—the gods of time, sun, earth, moon, Venus, as well as those of fertility and water. In a sense, 260 was the Maya divine temporal common denominator. What better way to make long-term projections than through the commensuration of all the divine cycles rolled together in the ultimate cosmic time unit?

number of days, and proceed directly to the beginning of the table. For example, one scheme calls for a termination of the count at the end of the 61st revolution of Venus in the table (four Venus periods short of completion of a great cycle). One drops four days, and then returns to the beginning of the table. This brilliant correction regimen had the following effect: first, it brought the canonic count into closer alignment with actual Venus appearances, and second, it resulted in a re-entry into the table on the appropriate ritual 1 Ahau date.[16]

Like Sosigenes and Gregory XIII, unnamed Maya chronologists repositioned an errant timekeeping scheme, and established a correction mechanism to reduce the drift between real and canonic time to proportions so tiny that they seemed to pass beyond practical motives. Also, as was the case for its Western counterpart, religion served as the driving force behind the architecture of the perfect timepiece. The 16th-century Spanish bishop and chronicler of Maya history, Diego de Landa, wrote that the priests in charge of religious ritual carried local versions of the codices from town to town. These documents contained predictions and ritual procedures pertaining to the worship of the celestial gods who managed the natural world.

Imagine one such priest-emissary (given what we explored in the last chapter I think it may be safe to call him an astrologer)[17] consulting his client, perhaps the chief of an allied city. He carefully opens the local version of the Venus Table in his codex and calculates the impending heliacal rise of the Venus deity. Might the dawning light anticipate inauspicious possibilities regarding the conduct of battles and captures? Perhaps the astrologer would signal the warning of a forthcoming eclipse at the next full moon, along with an omen related to this season's maize crop.

There is no escaping the conclusion: Maya astronomy, like that in much of human history, was driven by astrological concerns. The Venus god (shown in Plate 6) that figures in every page of the Venus Table at the middle right was a contact point between the essence of the science of astronomy and the woes of the common person, for the Maya believed their future was ordained in the stars. They would periodically renew their perceived cosmic connection by participating in carefully timed, astronomically based rituals. When Venus made its heliacal rising along the axis of a specially oriented temple in their city, the people would go there to witness a demonstration that their living, ruling lord was the very incarnation of his heavenly ancestors. Appearing before the altar bathed in the light of his patron star, the exalted ruler might perform a blood-letting ritual such as that described on p. 167. Thus, the ruler would demonstrate before the assembled masses the shared belief that he was the blood descendant of the sky gods who created the world and also held the power to destroy it. Except perhaps for the undocumented mental meanderings of the astronomer-scribe steeped in his profession, the Maya sky was less a source of objective knowledge, more a device for bonding people together.

Religious and astrological motives aside, I have always felt that mechanically elegant tabulations like the Dresden astronomical tables (there is another that pertains to eclipse prediction and a third that follows the course of Mars) offer us a glimpse of the pinnacle of achievement of the ancient Maya skywatchers. Here were astronomers who surely were as fascinated with timekeeping as the Old World computists—and indeed as are we with the mysteries of the unknown. And they were equally driven to discover new insights into solving those mysteries. I wonder, when we look at the codices, whether we might be witnessing the prodigious output of an elite, esoteric "Institute of Advanced Astronomical Studies."

Why elite? There is no doubt that the astronomers who observed sky events and interpreted their portents, as well as those who recorded, recopied, and updated the documents, were a part of the ruling class. Early Spanish chroniclers tell us that exalted Maya daykeepers taught their "sciences and ceremonies" to the children of royalty, that they were showered with gifts, and that their nearest relatives inherited their duties.[18] But was the scribe perhaps destined by his innate talents and skills to rise up the ladder of success to his exalted position, or was he a member of the royal bloodline simply groomed for the job? We don't know.

We began this chapter with the assumption that people are compelled in their search for order in nature to mark events and place them on a scale. We say they arrive at a calendar once they devise and express a method (usually a computational one) for parceling out time, thus enabling them to reach future dates. In the two examples I discussed—one from Roman, the other from Maya society—the ultimate goal of the calendar computists seems to have been to make their methodology more and more perfect by correcting the observed deficiencies. We can appreciate the difficulty of undertaking such goals by civilizations with limited technology only by exploring timekeeping systems in all their intricate detail—hence this chapter in *People and the Sky* that is a bit more (but hopefully not too) arduous in its complexity than the others.

We have followed only isolated fragments in the development of the calendars of two widely separated, unrelated cultures, one canon involving the sun and the moon, the other pertaining to Venus. The first part of the discussion will tie in with the pursuit of astronomy in the West, which is the subject of my final chapter. In each calendar the principal problem lay in setting ritual dates. The integral incompatibility of natural periods was, in both instances, the chief impediment to the solution. Once each calendar was put into operation for an extended period of time, slight imperfections arose. Attention was then turned to the delicate task of attempting to correct the slippage between natural and canonic time. At this point, the domain of relevant parameters and restrictions began to expand well beyond purely astronomical concerns. And yet our will to master time persists!

CHAPTER 11

The Western Sky

Development of Western Science is based on two great achievements, the invention of the formal logical system (in Euclidean geometry) by the Greek philosophers and the discovery of the possibility to find out causal relationship by systematic experiment (Renaissance).[1]

Albert Einstein

Modern experimental science is a child of the European Renaissance, with roots traceable both to the ancient Greeks, who gave us the gifts of logic and reason in the form of geometry and geometrical modeling, and to the Babylonians, whose arithmetical way of calculating was passed on to the Greeks. While Western science exhibits some of the hallmarks of many of the other cultures' ways of knowing nature, it is also characterized by a number of differences. How and why the West is different is the subject of this chapter. To accomplish our goal we need to explore each of these contributions.

Due to its geological and meteorological conditions, it is not surprising that writing in clay was invented in Mesopotamia. Characters there consisted of curious juxtapositions of triangular impressions we call "cuneiform" (cunei meaning wedge-shaped), hammered onto wet viscous earth compounds that hardened into almost eternal permanence. These clay tablets became the logical medium to carry astronomical as well as other information. Recall our discussion (in Chapter 1, "The Storyteller's Sky") of the Babylonian creation myth,

Enûma Elish, which appeared on seven huge cuneiform tablets that once adorned the city center of 7th-century BC Babylon.

Greasy clays were abundant all over the land between the Tigris and Euphrates Rivers. After the mud left by the annual flooding dried and caked, its components created a natural medium for visible symbols that could be systematically arranged to record past events. Unlike Maya writing, which seems to have been invented to record dynastic history and rituals pertaining to divination, writing in the Middle East rose out of a need for accurate bookkeeping (Fig. 45). Most cuneiform texts that survive deal with quantities of trade items. There are bills of lading, even recipes for making beer. But tablets also have been excavated that tell us about Mesopotamian law, and stories of creation.

One of the earliest examples of astronomical Middle Eastern writing appears in a cuneiform tablet named after a king who ruled in about the 17th century BC. The Tablet of Ammisaduqa specifically tells

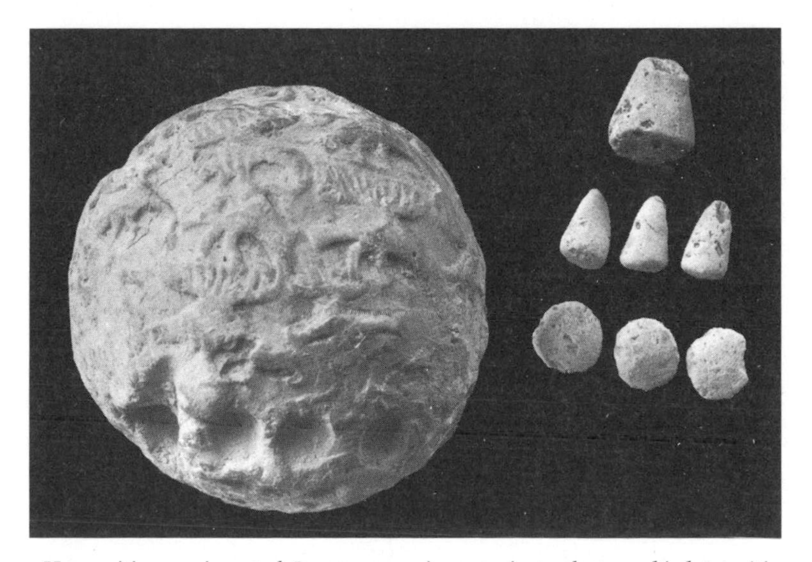

45 *How writing was invented: It was economics, not science, that gave birth to writing. Cuneiform developed out of stamped impressions made on clay envelopes of various shapes of tokens included within. Each shape stood for a trade item and the number of stamps or tokens for the quantity of each item.*

us about the kinds of records Babylonian astronomers kept and how they used their observations to make celestial predictions. An excerpt from it reads (note the familiar "when … then" notation):

> If on the 6th day of Ab Ishtar appeared in the east, there will be rains in heaven; there will be disaster. Until the 10th of Nisan she will stand in the east; on the 11th of Nisan she will disappear, and, having remained absent 3 months in the sky, on the 11th of Tammuz Ishtar will shine forth in the west; hostilities will be in the land; the harvest of the land will be successful.[2]

This subject of the text is the planet Venus (Ishtar). The scheme for marking time may be familiar to those of us who pay some attention to what happens to planets in the sky. We encountered something like it in our discussion of the Maya Venus Table in the Dresden Codex in the previous chapter.[3] Thus, the planet Venus is said to disappear on a particular day of a given month and to return on another. The observations refer to the times when Venus disappears in the light of the sun, after having been the evening star for several months, and when it reappears as morning star following a brief absence of several days—and the converse. Like their Maya counterparts, Babylonian astronomers tabulated intervals between each of these stations. Similar texts chart the phases of the moon, eclipses, dates of solstices and equinoxes, and so on. But make no mistake—all are concerned with predicting future sky phenomena for the sake of extracting omens from the sky gods.

Mesopotamian astronomical tablets mark stations in space as well as time. Mars, for example, is said to stand first in one degree of the zodiac and then in another. The interval between stations gives the degrees of separation. In one tablet, astronomers record a series of lunar eclipses and then use these data to calculate the difference between the times of occurrence of future eclipses. Their arithmetical way of forecasting is based on a simple formula:

PRESENT PLACE + SPATIAL INTERVAL = FUTURE SPACE
or
PRESENT TIME + TEMPORAL INTERVAL = FUTURE TIME

The first scheme charts the future place in the sky where an event might be expected to be observed, while the second gives the time in the future when an event is forecast to take place.[4] The Box: *Moon and Sun Positions in a Cuneiform Tablet* (p. 216), which deals with the prediction of new moons, offers a brief lesson based on data from an actual cuneiform tablet. It will help readers understand how this form of predictive astronomy, one of the major staples inherited by modern science, worked in practice.

Of what practical use is such a table? For the Babylonians it would have had immense value. They counted the days of the month from the first visible lunar crescent in the west, which followed conjunction by one or two days. Their business affairs, the setting of religious holidays, their entire schedule of livelihood depended on "the days of the moon" (if you make a car or rent payment on the first or last of the month, you can readily sympathize with the need to anticipate when the event that commenced the short cycle in their calendar would have taken place).

Modern scientific prediction works in much the same way. We express what we see in nature in the language of mathematics—we call these quantitative expressions the "universal laws of nature." Once we insert time into them, the equations generate specific predictions about the future. Testing the predictions by observation, we use a feedback process to alter and modify the equations. We add terms and insert factors into them to yield more accurate predictions that once again can be validated in the world of experience, and so on. Sometimes we detect even more subtle patterns in our observations. This can cause us to overturn some of nature's old laws in favor of new ones.

Our later (6th to 5th-century BC) Babylonian ancestors were particularly keen about designing arithmetical schemes that demonstrated how the movement of the planets fit together harmoniously. Here is one problem that preoccupied them: How often will a meeting of planetary deities (a conjunction of planets) take place? An astronomer's diary from this period reads:

Moon and Sun Positions in a Cuneiform Tablet

This translated excerpt from a Babylonian tablet[6] was written in the year 3, 27 (in sexagesimal notation[7] this means the 3 × 60 + 27 or the 207th year of the Seleucid era, or 104 BC in our calendar). The text consists of a listing by months of the year (Roman numerals, in the first column, beginning with the last (12th) month of the previous year) of the *anticipated* positions of conjunctions (close passages) of the moon and the sun in the zodiac (the appropriate constellation appears in the right-most column).

Take for example the first position (**Place**), in the constellation of Aries, expressed in degrees from zero to 30. It is written 2, 2, 6, 20, (third column) or 2° 02′06″20‴ in the table. Add to it the interval in the second line, 28° 50′39″18‴ (second column):

$$
\begin{array}{r}
2^\circ\ 02'06''20''' \\
+\ \underline{28^\circ\ 50'39''18'''} \\
30^\circ\ 52'45''38'''
\end{array}
$$

(Note: the symbols ′, ″, and ‴ stand respectively for the 60 divisions of the degree in the sexagesimal system that we call minutes, the 60 divisions of the minute being seconds, and the 60 divisions of the second, rarely seen in today's geometry texts. Incidentally, the intervals vary because the moon does not move at a uniform speed among the stars.)

Since each zodiacal constellation occupies 30° of the ecliptic (there are twelve of them), to determine, for example, how many degrees into

> ... Dilbat [Venus] 8 years behind come back ... 4 days thou shalt subtract ... the phenomena of Zalbatanu [Mars] 47 years ... 12 days more ... shalt thou observe ... the phenomena of Kaksidi [Sirius] 27 years ... come back day for day shalt thou observe ...[5]

Here for the first time we witness both the arithmetical notation discussed above as well as the habit of seeking commensuration, or the

TRANSCRIPTION OF PART OF A LUNI-SOLAR EPHEMERIS FOR THE YEAR 207 OF THE SELEUCID ERA[8]

Month	Interval	Place	Constellation
XII	29,8,39,18	2,2,6,20	Aries
208 I	28,50,39,18	52,45,38	Taurus
II	28,32,39,18	29,25,24,58	Taurus
III	28,14,39,18	27,40,4,14	Gemini
IV	28,24,40,2	26,4,44,10	Cancer
V	28,42,40,2	24,47,24,18	Leo
VI	29, ,40,2	23,48,4,20	Virgo
VIa	29,18,40,2	23,6,44,22	Libra
VII	29,36,40,2	22,43,24,24	Scorpio
VIII	29,54,40,2	22,38,4,26	Sagittarius
IX	29,51,17,58	22,29,22,24	Capricorn
X	29,33,17,58	22,2,40,22	Aquarius
XI	29,15,17,58	21,17,58,20	Pisces

Taurus, the constellation that follows Aries in the zodiac, the moon has progressed, we need to subtract 30° from the total. This yields the place where one would expect the conjunction to occur. This is 0°, 52′, 45″, 38‴ (written 52, 45, 38 in the table second place in the third column), and so on throughout the remainder of the text.

fitting together of diverse periodic cycles (discussed in the previous chapter), brought together in a single statement. The ancient astronomer's words offer a formula for approximating the observable cycles of the planets by adding or subtracting a small number of days from a full cycle.

The first part of the statement is obvious. It says that if you want to peg a Venus event in this particular year you need to "come back" four days every eight years. This is because, as we learned in the last chapter, a whole number of Venus cycles (five of them) comes closest to fitting our year, overshooting a whole number of years by about four days (2.3 days by modern measurements). For Mars, 47 of our years and "twelve days more" is a somewhat less approximate formula for commensurating a whole number of Mars cycles and seasonal years (it turns out to be 22 days). The last part of the statement remains a mystery. It seems to involve an attempt to tie all of these commensurations to the year cycle reckoned by the movement of the sun among the stars. The reference point may be a pair of heliacal risings (or settings) of the brightest star in the sky, Sirius, 27 years apart. To understand what we mean by Venus and Mars cycles, see the second Box: *Synodic Periods of the Planets*, p. 220.

Once they had collected all the observed phenomena associated with a particular planet and laid them out cyclically (e.g. first appearance, greatest angular distance from the sun, last disappearance), astronomers were able to compute that planet's "stations" for any run of years in the future by simply recopying and setting in place each of the cyclic phenomena within that period of years, and adding or subtracting the tiny corrections: four days for Venus, twelve for Mars, and so on. With this information they created an ephemeris of extraordinary value, for it could tell them precisely when and in which constellation future events would take place. A more long-range set of predictions, such as when planets would arrive at conjunctions with one another, could also be generated by such a table. For example, recall that close conjunctions of Jupiter and Saturn in particular captivated Chinese astronomers; they happen about every twenty years and they recur against the same stellar background every 960 years.

Long interval formulae such as these give us an impression that ancient astronomers in the West were deep thinkers, whose ruminations extended to the limits of the esoteric contemplation of time. This may have been the case, though unfortunately little evidence survives that deals with their philosophical pursuits. Above all, the work of the ancient astronomer had practical value. One clue that these numbers and arithmetical formulae pertaining to what goes on in the sky were more than contemplation and curiosity lies at the surface of practically every passage they wrote. Take the Tablet of Ammisaduqa. The subjective terms, such as "hostilities," "happy," and "successful" that appear in it, would certainly seem strange to us if we encountered them in a modern astronomy text. But let's not allow the omen-bearing qualities our Babylonian ancestors assigned to their astronomical tables to dissuade us from being interested in them, for these perspicacious skywatchers were indeed the founders of the modern science of astronomy. They needed to know the exact place of every moving denizen of heaven in order to anticipate the will of the gods. For them each planet was a deity whose whims and actions were thought to affect all who resided below. Stated simply: their astronomers, like ours, were scientists—but those who would follow them begin to look just a bit more like our modern scientists.

For most of the history of Greece, belief in a universe controlled by the gods was essentially the same as it was for the Babylonians, from whom they borrowed extensively. Greece, too, had its skilled astrologers. Their mandate was to "save the phenomena"; that is, to formulate a mathematical mechanism capable of generating future times and positions of celestial bodies for the primary purpose of extracting omens. As we learned in Chapter 9, Greek astronomer-

Synodic Periods of the Planets

When we speak of a planetary cycle we usually mean its synodic period. The "synodic" period (from the Greek word *synodos* or "coming together") is the interval between successive repeatable aspects of a planet with respect to the sun. It is directly observable. For example, suppose on a particular day Venus makes its heliacal rise, or first pre-dawn appearance, and is visible only for a minute or so following after an interval during which it has been lost in the glare of the rising sun.

Let us mark such a hypothetical date, say 1 July 2008. On the next day Venus will be seen for a few more minutes. As it stretches farther from the sun in the weeks that follow, Venus becomes visible for many more minutes, then for a few hours before dawn, until, after a few months, it reaches its maximum distance from the sun in the sky; then it slowly begins its return. Falling toward the rising sun, a few months later it vanishes in the morning light, becoming invisible for several more weeks. Then Venus reappears in the evening twilight, hovering over the sunset position for but a few moments before disappearing below the western horizon as it follows the sun. In the next several months Venus repeats the track it has outlined in the morning sky. Now, as evening star, it stands high in the west before returning, like a yo-yo on the end of its string, to another encounter with the bright luminary. Gone again, this time for a brief period averaging eight days, it finally returns to morning heliacal rise to complete its synodic cycle.

The synodic period of Venus is the sum of four distinct intervals, two of appearance and two of disappearance, totalling 584 days on the average. Thus, the date of the next morning heliacal rise would take place on 1 July plus 584 days, or 5 February 2010. Because five of these heliacal rises is nearly equal to eight of our seasonal years, the fifth heliacal rise, timed from 1 July, would happen on (nearly) the same date eight years later, or 29 June 2016 to be exact. Since most human activities are keyed to seasonal

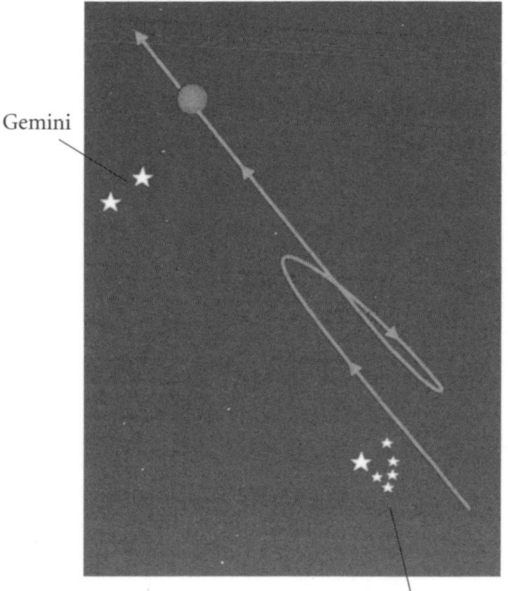

As the planet Mars passes eastward from Taurus into Gemini, it gradually slows to a halt, reverses its direction for several weeks, then resumes its normal west-to-east motion through the zodiac. The two turning points, to which astrologers all over the world paid particular attention, are called first and second stationary. This curious, looped path beguiled astronomers from ancient times to the Renaissance. Today we think of retrograde motion as a phenomenon that takes place when the earth passes Mars as the two orbit the sun. Then Mars seems to go backward against the background of distant stars for a while.

dates, this sort of celestial harmony, envisioned perhaps through recognition of a coincidence of Venus' heliacal rising events with the start of the rainy season, would serve as a logical basis to establish a calendrical reference point.

By contrast, Mars' synodic period is 780 days and, unlike the case of Venus (and Mercury), its synodic period (along with that of the remaining visible planets (Jupiter and Saturn) consists of a single lengthy appearance followed by a single brief disappearance period (See Chapter 1). Once Mars makes its heliacal rise in the evening dusk it wheels away from the setting sun, moving all the way around the night sky until it rejoins the sun at the opposite horizon; then it disappears in the pre-dawn light. Halfway through its course, when it lies directly opposite the sun, it makes its curious backward turn, called retrograde motion. While Venus repeats its moves in 584 days, Mars does so in 780 days.

astrologers were the ones responsible for the basic forms of the horoscope whose vestiges we still find online and printed in today's newspapers. But the Greeks also bequeathed us the concept of theorizing and speculating on our own existence and questioning with skepticism—qualities we tend to think of (again with little proof) as unique to Western thought. Nonetheless, as far as we can tell, the logic and reason we apply in our modern scientific way of acquiring knowledge about the universe is likely unique to the Greeks. It is perhaps their greatest gift to us.

The Greeks invented a very special form of logic—a space-based way of reasoning through celestial behavior patterns. Though they developed it some 2,500 years ago, it still serves as the foundation of the way we have come to understand the universe. We call it geometry, from *geometrien*, or "to measure land" as they referred to it. Today we conceive of the universe we live in is a complex of orbits confined to planes arranged in both flat and warped space, primarily because we inherited this way of thinking from them. Expressed in material form as a working mechanism with interconnected moving parts, they called it an *eidolon* in Greek or *simulacrum* in Latin (whence our word simulation) that describes how natural phenomena behave. The *Antikythera* mechanism, shown in Fig. 46, is an excellent example.

The Greeks seemed to be after a deeper truth, one they believed lay embedded within the tangible world. They believed they could discover the underlying principles that governed heavenly motion if they looked beyond the gods. Theory, logic, models—all are part of the Greek heritage, cast indelibly onto the pages of every modern scientific text we read about the natural world. To grasp the taproots of Western skywatching we need to open each gift package from the past and examine its contents carefully. We also need to ask: have these gifts remained pristine since their transmission to us over the ages? If not, how have the contents been tarnished or altered? We begin with the idea of making models and showing how they employed geometry.

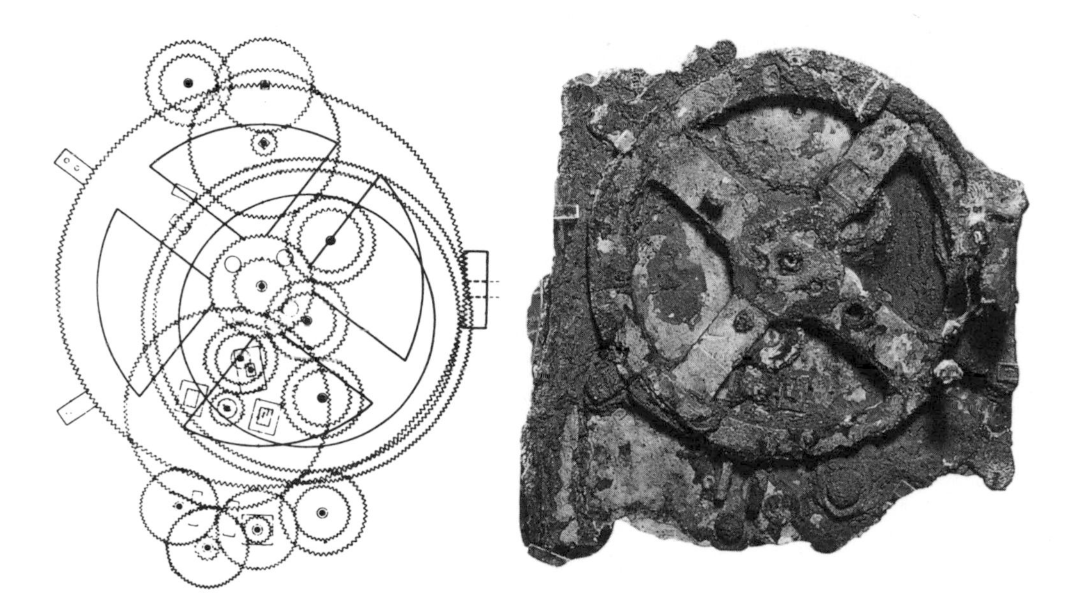

46 *The "Antikythera" Mechanism: Is it the first machine model of the universe? Early in the 20th century, a diver in the Aegean Sea brought up a mysterious mechanism from a wreck dated* AD *87 near the island of Antikythera. When the device was examined, it was possible to see an inner assembly of toothed gears that seemed to harbor astronomical data. It turns out that 19 rotations of the main wheel on the Antikythera mechanism coincided precisely with 235 rotations on another, just as 19 seasonal years fit precisely into 235 lunar synodic months—commensuration in material form par excellence!*

Early Greek models of the universe were based on common sense. For example, Thales of Miletus (624–546 BC) postulated that the earth is a flat disc surrounded on all sides by water (Fig. 47). He reasoned that it must be a disk because that is what the horizon that circles around us seems to look like. And it must float on water because we can witness ground water surging up from below in the form of artesian wells. That there must be water above us is proven every time rain falls. And as far as traders and explorers could make out, the horizontal limits of the earth disk were encompassed by a vast ocean-sea. Thales' water-bound terrestrial model is both practical and logical. It derives its explanatory power from everyday experience.

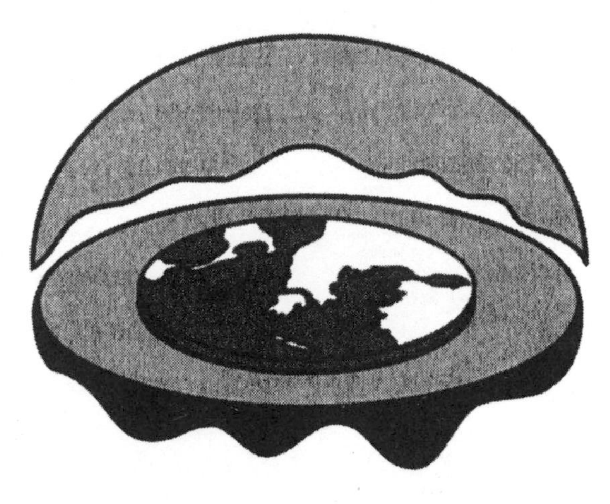

47 *Gifts from the Greeks: The flat disc of the earth lies at the centre of Thales' waterbound cosmos (550 BC).*

Anaxagoras' (500–428 BC) world model focused more on the heavens, but it was just as practically minded. He conceived of a cylindrically shaped geometrical world (we live on its flat-topped surface) that floats freely in space. The stars are appended to a sphere that rotates around the cylinder and carries them below the cylinder every night, then back into view the next day. The moon shines by the light of the sun, for if we watch its phases we can see for ourselves that the lighted portion of the lunar disk always faces the sun. Eclipses of the moon happen when the earth's shadow falls upon it. You can tell that because whenever an eclipse takes place we here on earth always seem to be located in the middle of the space along a straight line between the sun and the moon. The common denominator between the geometrical models of Thales and Anaxagoras is that simple observations of nature coupled with basic common sense serve as their backdrop.

No hidden forces, no abstract qualities—in these models everything makes sense to anyone who interacts up close with the natural environment on a daily basis. You get a picture of the world exactly as you experience it. Why this attitude? These philosophers were reared in the free-thinking, highly individualistic social environment of the Greek colonies of 6th-century BC Asia Minor and southern Italy. With power decentralized from the priestly hierarchy in the Peloponnese, there was no need to dress accounts of the workings of nature in religious garb. It is easy to understand why Greek models of how things work made a deep impression on the architects of modern science who lived in the time of the similarly oriented, free-thinking European Renaissance.

Another of the traits of modern astronomy rooted in ancient Greece is the habit of theorizing—and it is anything but practical. Fifth-century BC Socratic philosophers came to view nature as a set of interrelated phenomena meant to be dealt with at an intellectual level; that is, to be contemplated by the mind. They acquired an abiding faith in a universe that operated on certain underlying principles. "Underlying" is the key word here, for the Greek thinkers of the Classical period, like our contemporary scientists, believed that what we see on the surface of things in the real world is only an approximation to a deeper hidden truth that exists in the form of unvarying principles that govern an imagined, ideal world.

For example, when, in the early 18th century, Sir Isaac Newton explained that a cart rolling over a flat surface ought to maintain a constant velocity forever, never coming to rest, he was expressing the theory that, in the friction-free universe that underlies the imperfect one we all experience, there would be nothing to interfere with a body's movement. In such a universe the natural state of things would be eternal motion in a straight line rather than the state of rest all things seem to arrive at in the superficial world that we experience. Newton reasoned that the friction produced by the microscopic bumps on the roadways of the perceived world only deceives the senses. It hinders us from perceiving a more fundamental, simpler world—an underlying universe free of friction. Newton's outlook was Greek!

Today's science shares with the pagan philosophers the faith that nature's deep-seated invariant principles ultimately account for why things happen the way they do. We have rounded out Anaxagoras' cylinder and set it into motion about the sun. We have transformed Thales' static water-bound earth model into a time-dependent precipitation-evaporation cycle. Still we believe, as firmly as he did, that it rains not because the gods will it to do so or because plants here on earth need water to grow, but when it is necessary. God-free theories based on unchanging sets of principles, argued rationally and logically, lie at the foundation of Greek science.

It should be obvious that once we shift the focus of our discussion of the roots of astronomy in the West from the Babylonians to the Greeks, we acquire a kind of astronomy that seems more comfortable to us. Imagining celestial bodies as spheres moving in a void, or appended to spherical shells rotating about a common center—this was a characteristically Greek way of expressing how the universe really *is*. The space-based, geometrical way of perceiving the cosmos lives on today in modern maps of the solar system and the Milky Way, and in diagrams explaining the Big Bang origin of the universe. It is the source of our concept of space travel, and we have extended it in the opposite direction to the microcosmic world of subatomic structure. Would the orbital model of the atom ever have been contrived had it not been for the Greek model of the solar system?

Reductionism is another Greek gift to contemporary science. In astronomy it is manifested by the concept of reducing to a minimum the sum total of observed motions of sky objects necessary to fit a working model that consists of the simplest aggregate of circles and spheres inscribed into three-dimensional space. We can best understand this novel Western concept by discussing the three basic celestial motions that preoccupied the Greek astronomers:

1 the relatively rapid east-to-west motion of everything in the sky in 24 hours;

2 the much slower west-to-east motion of the moon, the sun, and the planets reflected against the background of constellations, each in its own period; and finally,

3 for the planets only, the periodic short-term reversal of this second type of motion during which the planet slows to a halt, turns westward for a time, stops again and resumes its normal eastward course among the stars. This so-called retrograde motion, which results in a curious loop for each planet if plotted out against the stars (as shown in the Box: *Synodic Periods of the Planets*, see p. 220), seems most beguiling of all. Even by the time of the scientific Renaissance in the 16th century, it continued to baffle both Copernicus and Kepler.

Creating models that reduced these motions to their simplest form was a challenge said to have been issued by no less pensive an individual than Plato. Eudoxus of Cnidus (408–355 BC) was one philosopher who aspired to the challenge. First to offer a theoretical explanation of planetary motion, Eudoxus devised a model that consisted of a set of concentric crystalline spheres, one assigned to each of the seven moving bodies (Fig. 48 depicts that of the sun). Each sphere rotated at such a rate and had its axis inclined at such an angle so as to "save the phenomena," that is, to geometrically reproduce the observed motion of each planet appended to it. The stars were positioned in the outermost sphere, to which all the others were attached. It turned about the earth-fixed center of the universe in 24 hours.

While Eudoxus' model did a pretty fair job of returning all things to approximately the correct place, it failed to account for the variable rate of movement of the sun on the plane of the sky over the course of the year; also, it did not explain why a planet's given retrograde loop differed slightly from the previous one. Only by making his model more complex—namely by adding additional spheres—could

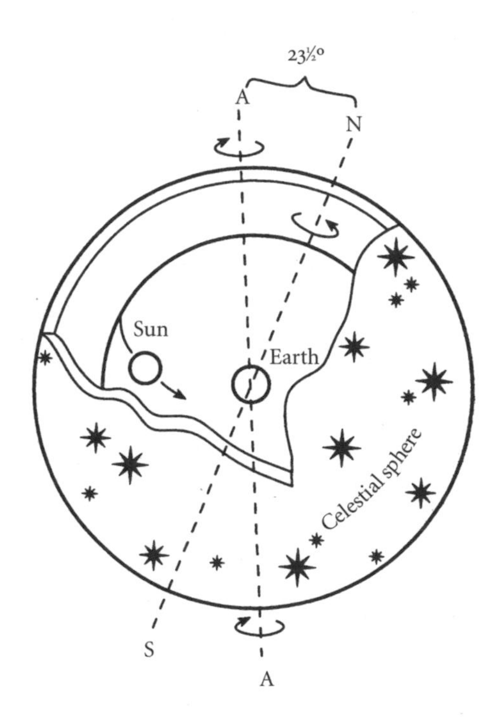

48 *Eudoxus' crystalline spheres (375 BC). The stars spin about axis N in a day while the sun turns in the opposite direction about axis A in a year. Each planet has its own set of spheres. Again the earth is at the center.*

Eudoxus' successors even begin to account for the subtleties of what once were perceived to be rather simple movements, but over the years had become more complex in the wash of data gleaned from more careful observation. At one point in the modification of Eudoxus' model, Mars, always a bit more errant than the other planets, itself required a total of seven spheres to save its phenomena.

Ingenious as it was, the spherical shell model of Eudoxus failed the test of reductionism—it had too many spheres! Clearly no gods, whether they meddled directly or indirectly in human affairs, would have created a universe as complicated as Eudoxus had tried to model.

Claudius Ptolemy, the Alexandrian astronomer of the 2nd century AD, devised a more successful model (Fig. 49). It retained the earth at the center of a spherical universe, but handled retrograde motion in a more elegant way by imagining each planet to orbit on an "epicycle," the center of which moved on a bigger orbit called a "deferent." Every time a planet, say Mars, reached the inner part of its epicycle and aligned with the earth, it appeared to execute a brief backward loop—retrograde motion. This effect can be compared to watching a piece of chewing gum stuck to a moving bicycle wheel. As the rider passes by, fix your eyes on the path of the gum. When it comes in contact with the ground and the wheel rolls over it, the gum momentarily reverses its direction, then pitches forward again as it rolls toward the top of the wheel.

Not only was Ptolemy's model highly imaginative, but also it reduced the number of spheres necessary to account for the three fundamental kinds of celestial motion—until more accurate observations revealed inconsistencies in his model. Attempts to correct the model by placing the earth slightly off center sufficed to save the phenomenon for a while. It would take nearly fifteen centuries before Copernicus overturned the Ptolemaic model by creating his revolutionary heliocentric (sun-centered) model of the solar system.

Many modern textbooks label Ptolemy's idea of a geocentric universe as "wrong" because the earth really isn't the center of the universe, or because a planet can't travel on an epicycle since there is no material body to attract it and hold it in its orbit. These conclusions stem from the habit of telescoping our notions and ideas about

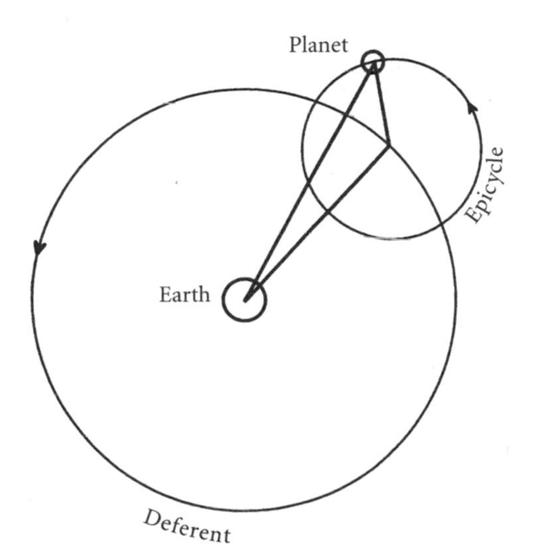

49 *The Ptolemaic Model (AD 150). In this geocentric model planets move on epicycles, the centers of which travel on deferents.*

nature across a vast sea of time and cultural change, and implanting them firmly in the heads of our forebears. For example, the concept of gravitation and the idea that forces can act at a distance along lines interconnecting celestial bodies was not devised until the time of Newton, who lived fifteen centuries after Ptolemy.

Seeking common ground can be more meaningful than searching for differences that set us apart from others who have looked at the sky. We ought to see Ptolemy's geocentric model of the solar system for what it was—a novel attempt to save the phenomena. Greek scientific skills were grounded in rationalism as well as in the ability to modify a theory in the face of new information. They were erecting the universe they desired, one controlled by hidden principles in the hands of gods who had far more important celestial business to tend to than to meddle directly in human affairs.

The ancient Greek models of the universe that we have been discussing are as mechanical and machine-like as the ones our scientists devise today. We say that the brain is like a sponge, the nervous system is like a computer, and the atom operates like its macroscopic counterpart, the solar system, as a kind of celestial billiard game. The Greeks may be gone but their ideas live on in our contemporary way of understanding the world.

EPILOGUE

Distancing the Sky

astronomy: "as·tron·o·my … the science that treats of the celestial bodies, of their positions, magnitudes, motions, distances, constitution, physical condition, mutual relations, history, and destiny—formerly used as synonymous with astrology"[1]

Webster's Dictionary

Why didn't they use fractions? How come they didn't know the world was round? Or that the earth orbits the sun? How is it that they didn't know about galaxies and the expanding universe—or care about life on other planets? What about supernova explosions? And the (26,000-year cycle of) precession of the equinoxes? What about the alphabet? No wheel? And how could they align their pyramids so accurately?

If you've read the first ten chapters of *People and the Sky* and contrasted them with the last, you now know the answers to the most common questions people have asked me over the years about astronomy in other cultures: *Because they are not us.*

Recall from Chapter 1 that our modern-day story is a monologue about stars, while theirs is a dialogue between stars and people. Ours has no blood, no sacrifice, no ancestral gods; theirs has no abstract laws, no purposeless forces, no random events. Among all the different categories of people we encountered: ruler, farmer, city planner, and so forth, we were impressed with the "lococentric" nature of the worldview they shared—the cultural response to the vagaries of the immediate environment. This cosmic view, too frequently mistaken for arrogance (how could they think themselves the center of the universe?), arose out of the need to subsist in a sphere of sensate comprehension far different from our own.

Still, you say, eclipses happen, the moon goes through its phases and Venus rises and sets heliacally everywhere on earth. All rational beings chart their existence in a seemingly chaotic world they presume to be ordered. We all observe natural phenomena constantly and consistently—whether the sky event be the daily setting of the sun, a close encounter between two planetary wanderers, an eclipse, or the fluctuation of light intensity in a distant quasar. Given the desire to seek order and unity, and given basically the same celestial images gliding over the heads of Polynesian and Babylonian, Maya and Egyptian, why is it that organized societies, all of whom live life under a canopy of flashing moving lights that occupies half their visible field, interacted so differently with what they saw? Why so many different astronomies?

Different outlook; different astronomies. As we learned in Chapter 8, "The Ruler's Sky," it made perfect sense for the Maya or Babylonian king to attribute his source of power to the cosmos—especially if he wished to convey the idea that his power is as eternal as the pristine movement of the sun, moon, and stars and as strong as an earthquake. And without technology intervening between people and the sky, such an outlook would be all the more believable.

And yet, there are similarities. In Chapter 9 we discovered that the astrologer of old was as equally motivated as today's astronomer to quantitatively search out the underlying harmony in whose existence they both believed. The motive then was perhaps more immediate: to peek around the corner of time and into the future so that one could meditate upon a suitable course of action. Today, with the possible exception of foretelling an asteroid collision, astronomical prognostications about the percentage of dark matter in the universe or the rate of deceleration of the Big Bang contribute to shaping our ideology—our cosmic overview. Some of us will regard that ideology as bleak—a non-participatory one-way world view in which a totally objectified universe cannot be influenced by human action. But others will revel in the high-minded thought that a hunk of tissue called the human brain, small enough to hold in the hand, can create wisdom and insight that links us to a chain of being that runs the gamut from galaxies to DNA. And as we also learned in the astrology chapter, the eureka moments that lead to that high-minded transcendency are as characteristic of the shaman as they are of the scientist.

Despite these common denominators, is it reasonable to ask whether people from ancient China, south India, the Andes, the Amazon, and Mexico—all with their own diverse customs and histories—understood nature the way we do? Unlike our own, their pasts did not necessarily consist of feudal state, imperial expansion, an artistic Renaissance, a religious reformation and counter-reformation, the sudden and rapid flourishing of technology, and the growth and development of the state. And even if our basic histories were the same, could we dare anticipate that the outcome of the discourse between nature and culture—after centuries of the mixing and blending of different people and ideas, much of it in long-term isolation from the West—would produce the precise, quantitative astronomical science we know today?

I cannot give a sure answer to such a question. But we have learned that navigator, storyteller, shaman, warrior, farmer—all of them astronomers in a sense—created wonderfully diverse and imaginative ways of understanding and explaining the world around them, each in a context that made sense to *them*. This, in itself, is worth marveling at. Like the unity among gods and people that held their universe together, ours is based on a similar kind of faith—the belief that a unity exists among all the forces of nature. For example we believe that gravity, which we choose not to identify directly with a god, behaves exactly the same way regardless of whether we choose to express mathematically the movement of a falling apple to the surface of the earth or the movement of a star that travels on an orbit about the core of a galaxy billions of light years beyond the range of our most powerful telescopes.

To a deeply held belief in unity we must add a second article of faith—progress. We believe that the totality of knowledge that we can perceive about the universe will lead us toward a truth that lies behind what we see. But if we've learned our science well, we know that truth will never ultimately and unquestionably be attained; it can only be reached approximately and by degrees (asymptotically as the scientists would say). Modern astronomical science is destined to create not only better and better theories and models that will yield empirically testable predictions, but also—and here we place an extreme demand upon our faith—theories and models that must be capable of being disproved as well.

Building abstract mathematical models, creating explanations that reduce the behavior of things to the fewest components, testing and retesting theories of the universe to make their predictions more precise—these are the hallmarks of the science of astronomy derived in the Western tradition that we explored in the previous chapter, so eloquently summarized by Einstein in that chapter's epigraph.

As one modern astronomer asks after laying down the 21st-century cosmologists' account of the evolution of the universe: "Who gets to celebrate this cosmic view of life? Not the migrant farm worker. Not the sweatshop worker ... You need the luxury of time not spent on mere survival."[2] Our Micronesian sailor, !Kung hunter-gatherer, and Stonehenge farmer were far too preoccupied to theorize abstractly about the sky.

If all this gives you the image of today's astronomer as a surgeon who methodically probes a passive, inert body of knowledge with the goal of extracting progressive new insights, then you've hit upon the one significant difference between us and them—all the others we've encountered in a wide variety of contexts and cultures in *People and the Sky* who, like us, have looked upward and outward. In "Patterns in the Sky" (Chapter 2) we mused about animal figures in the Martian landscape, on cave walls, and in profiles of puffy clouds. We found that the eye and the imagination work together to create a diverse imagery. All constellations are patterns familiar to those whose imagination creates them, but their organization into zodiacs reflects a way to seek order—in this instance to organize the passage of the wandering sky deities or the seasons of the year into a pattern recognizable to all. Their astronomy was *lived* as much as it was practiced, and much of what our predecessors included in its domain served what we might regard as social and religious rather than scientific purposes. Yet, as we've seen, when we chip away at the bedrock of Western astronomy, we find myth, astrology, and other social and religious concerns buried in its substrate. Perhaps deep down we all believe (or would like to believe) we really are part of an animate world and that somehow we are capable of recovering our lost intimacy with the sky.

Astronomy was once part of a dialogue between everyday people and supernatural forces thought to be intimately tied to celestial phenomena. I think that is because both the ancient and contemporary astronomers in the ancient cultures we've encountered—from Maya to Mursi and from Babylonian to Barasana—experience nature close up, without a complex technology to intervene between sky and skywatcher. While modern astronomers use remote sensing to probe the sky, our ancestors were more inclined by experience and hence better equipped to see the heavens as animate, and therefore closely attached to every activity they undertook. Theirs was less a one-way probing and more a two-way conversation with vital cosmic forces intimately tied to celestial phenomena, and quite receptive to human action. It informed cultural world views, ideologies, ways of relating the individual to society, and nature to the transcendent.

If learning is about seeing something in a way you haven't seen it before, then let the lesson of *People and the Sky* be that astronomy isn't what you thought it was. The dictionary definition of astronomy quoted in my epigraph to this chapter simply doesn't cover all bases. It is too one-dimensional. It depends too much on recording celestial imagery in the form of writing and tabulations—the way we have all learned about since our school days to acquire and understand the record of the past. That's our astronomy—not all astronomy. In my view we would do better to think of astronomy as a *human* (rather than only a Western) inquiry into the nature of things celestial. I challenge my readers to formulate definitions of their own that make sense to *them.*

Notes

INTRODUCTION

1 Weinberg 1988: 159.
2 Jonnes 2003: 75.
3 Carter 2006. I am indebted to Michael Haines (Department of Economics, Colgate University) for providing this source.

CHAPTER 1

1 Sahagun 1953: 1.
2 Sahagun 1953: Book 7, 4.
3 *ibid.*
4 Sagan 1980: 205.
5 Tedlock, D. 1985: 72.
6 Tedlock, D. 1985: 159–60.
7 Heidel, *Enûma Elish* I, line 4.
8 *ibid.*: iv, 95–102.
9 *ibid.*: v, 1–22.
10 Zolbrod 1984: 90.
11 *ibid.*: 83.

CHAPTER 2

1 Brown 1900: i, 127ff. and i, 220–26.
2 Cf. e.g. Condos 1997 and Krupp 2000.
3 Krupp 2000: 9.
4 For a detailed discussion of this myth see Magaña 1988 and 1996.
5 Goodman 1992, whose interpretation I have employed, gives a full recounting of this tale.
6 Pannekoek 1961: 57.
7 Reiner and Pingree 1985: Part 2, 41.
8 Krupp 1978: 216–219 gives a full account.
9 See Bricker and Bricker 1992.
10 Hugh Jones 1982.
11 Urton 1981.
12 Urton 1981: 110.
13 Morieson 1996: 102.

CHAPTER 3

1 Aratus, *Phaenomena*, cited in Roy 1984: 181.
2 Forster 1778: 528.
3 Dening 1962: 111.
4 Rieu 1946: 95.
5 Pryde, in a letter to MacDonald: 167.
6 *ibid.*
7 *ibid.*: 169.
8 Gladwin 1970.
9 Henry 1928, cited in Kursh and Kreps (unpublished manuscript): 86–87.
10 Diamond 2005: 86.
11 See, for example, Finney 1994 and Lewis 1976. But not all who have studied the matter agree that this, in fact, *was* done. For a thorough discussion of the navigation controversy, see Finney 2005: esp. 340–1.

CHAPTER 4

1 As told to Gatty 1979: 44. Harold Gatty was an experienced navigator. He flew with Wiley Post on his record-breaking eight-day around the world trip in 1931. In the work cited he recounts methods for determining location and pathfinding from signs in nature.
2 Gatty 1979: 109.
3 Marshack 1989.
4 Murray 1982.
5 Silberbauer 1981: 111.
6 The Heelstone, which still stands, may have had a twin to comprise this gateway. Looking in the opposite direction, one encounters the horizontal Slaughter Stone near the center of the Trilithon Horseshoe, which, once erect, likewise may have possessed a companion, thus creating a southwest-facing gate through which one may have

viewed the midsummer full moon (or midwinter sun) set.

7 Evidence from the nearby Durrington Walls occupation site suggests a thriving society existed *c.* 2500 BC.

CHAPTER 5

1 Aaboe 1974: 21–2.
2 Dicks 1970: 37.
3 Frazer 1983, Hesiod: 382–87.
4 *ibid.*: 609–614.
5 *ibid.*: 414–419. The *Iliad* and *Odyssey* of Homer also allude to celestial timing, mentioning Sirius (which makes us suffer from the heat), Arcturus, the Pleiades, the Bear (our Big Dipper), who, Homer tells us, does not participate in bathing in the ocean.
6 Aveni and Ammerman 2001.
7 Psalms 89: 12.
8 Deuteronomy 11: 11.
9 Amos 9: 13.
10 Ezekiel 36: 8, cf. Cohn: Chapter 3.
11 Ammarell 2005.
12 The mathematics is as follows: l = h tan £, where £ = ° - ‰; ° = latitude; ‰ = declination of sun, h = the height of a vertical pole and l = the length of its shadow. We find, for the June solstice, lJS = 1.989 lDS, where DS represents the December solstice, and ° = 7°S, the mean latitude of Java. The extremes of latitude are 5°50´S to 8°30´S. (cf. the standard calendar in Ammarell 1996, Table 1).
13 For example, Hines 1991.
14 Aveni 1993b.
15 See Aveni 2003 for a brief history of the origin of our seasonal holidays.

CHAPTER 6

1 Louis 1975: 3.
2 Fletcher 1902: 732–3.
3 Louis 1975: 3. (Griffin-Pierce 1992: 94.)
4 For other interpretations, see Griffin-Pierce 1992, and Chamberlain 1992.
5 If you draw a line from Corona Borealis through the north celestial pole and extend that line the same distance, it will strike the Pleiades exactly.
6 Murie (in Chamberlain 1992: 230).
7 Makemson 1938: 377.
8 Fletcher 1902: 733.
9 As mentioned earlier (note 5), the two star groups are exactly equidistant from the pole. They really balance about the polar pivot like a perfectly weighted see-saw. The sky, say the Iroquois, must be "gendered for balance"—neither half can exist without the other. For an excellent discussion of the gendering of the sky, see Mann 2000: Chapter 2.
10 Mine is a simplified description. Actually there are six mixed subclasses and their marriage interrelationships are too complex to elaborate upon here. But see Johnson 1988: 69–76 for a discussion.
11 Rowe 1979: 27.
12 Maurtua 1906: 150–2.

CHAPTER 7

1 Sahagun 1961: 191–2.
2 Meyer 2001: 2, note 2: Universal Asylum and Columbian Magazine, March 1792.
3 Lopez Austin et al 1991: In ancient Mesoamerica, thirteen was regarded as the number of layers of heaven, nine that of the underworld; and eighteen was the number of months in a year.
4 The work was done by a team of archaeologists headed by Saburo Sugiyama. An excellent exposition of the work appears in the archaeology online website, archaeology.org.
5 Sahagun 1981: 42.
6 The Mind of God is actually the title of such a book (Davies, 1991).
7 Aveni and Romano 1994: One problem with interpreting the data is that some temples consist of two or three chambers, each dedicated to a different deity.

8 Thulin 1913: 134, 136.
9 Pankenier 1983: 512–3.
10 Aveni 1993a: 81.
11 For a detailed discussion of the plan, see Krupp 1989.
12 Needham 1974: 79.
13 Wilson 1988: 6.
14 For a more detailed discussion of the cosmography of the U.S. capital, see Meyer 2001, note 2. I thank R. Horn and J. Horn (p.c. 24 Feb 2005) for direction and discussion on this matter.

CHAPTER 8

1 Hocart 1941: 1.
2 Waterhouse 1633.
3 For an excellent modern account of the story of the attacks and their aftermath, see Rountree 2005, whose book I have consulted in preparing this chapter. I have also used material from my own piece (Aveni 2005) concerning how the attack was timed celestially.
4 Both attacks also occurred on Good Friday which, argue some historians, was chosen to symbolize a rejection of the English religion.
5 Strachey, cited in Miley Theobald 2005: 23–4.
6 Waterhouse 1633: cited in Aveni 2005: 29.
7 "When Anu had created the heavens...," from a 1st-millennium BC tablet found at Babylon (Heidel 1942: 65).
8 Pannekoek 1961: 91. For more on planetary lore, see Aveni 1993b.
9 Salomon 1991: Chapter 5.
10 Tedlock 1985: 146.
11 Krupp 1997: 167–173.
12 *ibid.*: 171.
13 For more on associations between Jupiter (and Saturn) retrograde motion and dates on Maya monuments see Milbrath (in press).
14 Identical combinations on the 260 and 365 day "time cycles" occur at intervals equal to the lowest common multiple of 260 and 365. This is equal to 18,980 days or 52 years of 365 days (see Chapter 10 for details).
15 Plato, Cratylus (Cited in Hocart 1941: 31).

CHAPTER 9

1 Neugebauer and Parker 1969: 214–5.
2 One of the clearest discussions of these terms is provided by Turner 1972: 437–444, from whom I here borrow.
3 Aveni 1993b: Chapter 8.
4 The term *oracle* can refer either to the place or to the medium.
5 Ray 1981: 174–190.
6 *ibid.*: 180.
7 Or according to a recent account, ethylene gas, which still issues from a nearby geological fault (Broad 2006).
8 For example, see Price 1986.
9 *ibid.*: 147.
10 Chime Rada 1981: 22–3.
11 Varro, *De Lingua Latina* vii: 7 (cited in Aveni and Romano 1994).
12 R. Thompson 2: xx, no. 210.
13 Pannekoek 1961: 39.
14 Saliba 2006: 238.
15 Cited in Aveni 1993b: 84.
16 Schele and Grube 1997: 148.
17 B. Tedlock 1992: 53.
18 *ibid.*: 162.
19 Cf. e.g. A. Aveni ed. 1992: esp. Chapters 1 and 10.
20 Dorsey 1905: Chapter 6.
21 Schleiser 1987: 104.
22 *ibid.*: 105.
23 Goodman 1992a: 31.
24 D. Johnson 1998: 78.
25 Hoyle 1981: 43.
26 Davies 1991: 227.

CHAPTER 10

1 Betts 1999: 135.
2 Three of the four are named after deities. The curious exception is Aprilis, after *aper*, or boar: "The Month of the Boar."

3 Today the period is reckoned at 285 days, but it can extend to 295 days or more.
4 The early evolution of the Roman calendar is dealt with in detail in York 1986. See also Aveni 1987.
5 Thus: Canonic year 365d 6h 0m 0s, Tropical year 365d 5h 48m 46s, difference 11m 14s
6 The idea of using a *mean* moon which falls either a little ahead or a little behind the *true* moon has become common in western calendrical computations. Today, to avoid hours of unequal length, we use a mean sun instead of the apparent sun in the sky for keeping the time of day. Moreover, we keep the hour by a mean meridian which we share with those who live many miles east of west of us. (Likewise, 2,000 years ago the Chaldeans reckoned the positions of planets in the zodiac by linear functions that approximated the observed motion of the planet.) In the U.S., Eastern Standard Time, for example, is reckoned by the 75°W longitude meridian, Central Standard Time by the 90th meridian, etc.
7 Quoted in Moyer 1982: 144–150.
8 By agreeing to convert AD 4000, 8000, and 12,000 to common years, the fallback has now been reduced to one day in 20,000 years. Finally, at an Eastern Orthodox congress held in Constantinople in 1923, yet another rule with far-reaching consequences was adopted: century years divisible by 900 will be leap years only if the remainder is 200 or 600. The resulting calendar is accurate to one day in 44,000 years.
9 J. Thompson 1971: 62.
10 Closs 1977: 96.
11 A) 146 passages by a fixed first day of the tzolkin count (146×260^{d}); B) 65 heliacal rise events of Venus (65×584^{d}); C) 104 passages by a fixed day of the 365-day count (104×365^{d}).
12 $8 \times 365^{d}.2422 = 2921^{d}.9376 = 5 \times 583^{d}.92 + 2^{d}.3376$. The moon is involved as well: $99 \times 29^{d}.5306 = 2923^{d}.5295 = 5 \times 583^{d}.92 + 3^{d}.9294$.
13 Thus: 584 days × 13 lines per page × 5 pages = 37,960 days.
14 The coefficient (1) is one of thirteen numbers, the day name (Ahau) one of 20 in a pair of cycles, 20 and 13, that mesh together to form 260.
15 $2 \times 260^{d} = 3 \times 173^{d}.46 - 0^{d}.62 = 520^{d}$
16 Thus: $(61 \times 584^{d} + 4^{d}) - (61 \times 583^{d}.92) = 0^{d}.88$ instead of the $5^{d}.2$ error per uncorrected great cycle, if the correction were not made. The full correlation scheme is a bit more complicated. With successive corrections of eight, eight and four days upon each 104-year passage through the table, the Maya managed to reduce the error to one day in five centuries (for details, see Aveni 2001).
17 We cannot be absolutely certain that those who wrote the codices were the very ones who administered their omens to the people. Thus there may have been a distinction between the profession of timekeeper and astrologer.
18 Tozzer 1941: 27.

CHAPTER 11

1 From a letter written by Albert Einstein in 1953 in response to a question about what makes science unique. Quoted in D. Price 1971: 15.
2 Langdon, Fotheringham, & Schoch 1928: Chapter 10.
3 For a discussion of the striking similarities between the Babylonian Venus Tablet and the Maya Venus Table, see Aveni 1993b.
4 But again, this formulation is not unique. We find the same methodology in the Maya record as well. See Aveni 2005.
5 Pannekoek 1961: 55.

6 Aaboe 1974: 24.
7 Like the Maya *tzolkein* (Chapter 10), sexagesimal (essentially base-six) notation developed out of a special need to create a system to facilitate the computation of time, in this case parts of the 360-day year recognized by the ancient Sumerian culture. This is where we get our degrees-minutes-seconds notation, not to mention our 24-hour day. All units are divisible by six.
8 After O'Neil 1986. Such tables must have been based on years of continuous observation of which constellations of the zodiac could be sighted before the sunrise and after sunset at various seasons of the year (one can never actually *see* the sun against the background of stars of the constellation in which it resides).

EPILOGUE

1 Gove 1993: 136.
2 Tyson 2007: 27.

Bibliography

Aaboe, A., 1974. "Scientific Astronomy in Antiquity," in *The Place of Astronomy in the Ancient World,* ed. F. Hodson, Philosophical Transactions of the Royal Society. London, A 276, 21–42.

Akerblom, K., 1968. *Astronomy and Navigation in Polynesia and Micronesia.* Stockholm, Ethnografiska Museet.

Ammarell, G., 1988. "Sky Calendars of the Indo-Malay Archipelago," *Indonesia* 45: 84–104.

———, 2005. "The Planetarium and the Plough: Interpreting Star Calendars of Rural Java," in *Songs from the Sky: Indigenous Astronomical and Cosmological Traditions of the World,* ed. V. del Chamberlain, J. B. Carlson, and M. J. Young. Bognor Regis, Ocarina Books and Center for Archaeo-astronomy, College Park, Maryland, 320–35.

Aveni, A., 1981. "Archaeoastronomy," *Advances in Archaeological Method and Theory* 4: 1–81.

———, 1987. "Some Parallels in the Development of Maya and Roman Calendars," *Interciencia* 12: 108–15.

———, (ed.), 1992. *The Sky in Mayan Literature.* Oxford, Oxford University Press.

———, 1993a. *Ancient Astronomers.* Washington, Smithsonian Institution.

———, 1993b. *Conversing with the Planets: How Science and Myth Invented the Cosmos.* New York, Times.

———, 1996. *Behind the Crystal Ball: Magic, Science, and the Occult from Antiquity Through the New Age.* New York, Times.

———, 1997. *Stairways to the Stars: Skywatching in Three Great Ancient Cultures.* New York, Wiley.

———, 2001. *Skywatchers: A Revised and Updated Version of Skywatchers of Ancient Mexico.* Austin, University of Texas Press.

———, 2003. *The Book of the Year: A Brief History of Our Seasonal Holidays.* New York, Oxford University Press.

———, 2004. "Intervallic Structure and Cognate Almanacs in the Madrid and Dresden Codices," in *The Madrid Codex,* ed. G. Vail and A. Aveni. Boulder, University Press of Colorado, 147–70.

———, 2005. "How Was 'the Tyme Appointed'?" *Colonial Williamsburg Foundation.* Williamsburg, Virginia, 27–32.

Aveni, A. and A. Ammerman, 2001. "Early Greek Astronomy in the Oral Tradition and the Search for Archaeological Correlates," *Archaeoastronomy: Journal of Astronomy in Culture* 16: 83–97.

Aveni, A. and G. Romano, 1994. "Orientation and Etruscan Ritual," *Antiquity* 68: 545–63.

Aveni, A. and Y. Mizrachi, 1998. "The Geometry and Astronomy of Rujm el-Hiri, a Megalithic Site in the Southern Levant," *Journal of Field Archaeology* 25: 475–98.

Betts, J., 1999. "The Growth of Modern Timekeeping," in *The Story of Time,* ed. K. Lippincott. London, Merrell Holberton, 134–37.

Blier, S., 1987. *The Anatomy of Architecture: Ontology and Metaphor in Batammaliba Architectural Expression.* New York, Cambridge.

Bricker, H. and V. Bricker, 1993. "Zodiacal References in the Maya Codices," in *The Sky in Mayan Literature,* ed. A. Aveni. New York, Oxford University Press, 148–83.

Broad, W., 2006. *The Oracle: The Lost Secrets and Hidden Message of Ancient Delphi.* New York, Penguin.

Brown, R., 1900. *Researches into the Origin of the Primitive Constellations of the Greeks, Phoenicians, and Babylonians,* 2 vols. London, Williams and Norgate.

Carter, S. et al., 2006. *Historical Statistics of the United States: Earliest Times to the Present,* vol I, Part A "Population." New York, Cambridge University Press.

Chamberlain, V. del, 1982. *When Stars Came Down to Earth: Cosmology of the Skidi Pawnee Indians of North America.* Los Altos, California, Ballena Press, and Center for Archaeoastronomy, College Park, Maryland.

———, 1992. "The Chief and His Council," in *Earth and Sky: Visions of the Cosmos in Native American Folklore,* ed. R. Williamson and C. Farrer. Albuquerque, University of New Mexico Press, 221–55.

Chime Rada, Lama, 1981. "Tibet," in *Divination and Oracles,* ed. M. Loewe and C. Blacker. London, George Allen and Unwin, 3–37.

Clagett, M., 1995. *Egyptian Science,* 2 vols. Philadelphia, American Philosophical Society.

Cleal, R., K. Walker, and R. Montagne, 1995. *Stonehenge and Its Landscape: Twentieth-Century Excavations.* London, English Heritage.

Closs, M., 1977. "The Date Reaching Mechanism in the Venus Table in the Dresden Codex," in *Native American Astronomy,* ed. A. Aveni. Austin, University of Texas Press, 89–99.

Cobo, B., 1956. *Historia del Nuevo Mundo* (1653), Biblioteca de Autores Espanoles, 91–92, Madrid.

Condos, T., 1997. *Star Myths of the Greeks and Romans: A Sourcebook.* Grand Rapids, Michigan, Phanes.

Davies, P., 1991. *The Mind of God: The Scientific Basis for a Rational World.* New York, Touchstone.

Dening, G., 1962. "The Geographic Knowledge of the Polynesians and the Nature of Inter-Island Contact," in *Polynesian Navigation,* ed. J. Golson. Wellington, Polynesian Society, 102–31.

Diamond, J., 2005. *Collapse: How Societies Choose to Fail or Succeed.* New York, Viking; London, Penguin.

Dicks, D. R., 1970. *Early Greek Astronomy to Aristotle.* London, Thames & Hudson.

Dilke, O., 1971. *The Roman Land Surveyors: An Introduction to the Agrimensores.* London, Newton Abbot.

Dorsey, G., 1905. "The Cheyenne: I. Ceremonial Organization," *Field Columbian Museum Publ 19, Anthrop. Series* 9, No. 1.

Finney, B., 1994. *Voyage of Rediscovery: A Cultural Odyssey through Polynesia.* Berkeley, University of California Press.

———, 2005. "Applied Ethnoastronomy: Navigating by the Stars Across the Pacific," in *Songs from the Sky: Indigenous Astronomical and Cosmological Traditions of the World,* ed. V. del Chamberlain, J. Carlson, and M. J. Young. Austin, University of Texas Press, and Bognor Regis, Ocarina Books, 336–47.

Fletcher, A., 1902. "Star Cult Among the Pawnee—A Preliminary Report," *American Anthropologist* 4: 730–36.

Forster, J. P., 1778. *Observations Made During a Voyage Round the World.* London, G. Roberston.

Frazer, R., 1983. *The Poems of Hesiod.* Norman, Oklahoma, University of Oklahoma Press.

Frigerio, P., 1933. *Antichi Istrumenti Technici: a proposito di una stela funeria romana del Museo di Como.* Como, C. Nani.

Gatty, H., 1979. *Nature is Your Guide: How to Find Your Way on Land and Sea.* New York, Dover.

Gingerich, O., 1987. "Zoomorphic Astrolabes and the Introduction of Arabic Star Names into Europe," *New York Academy of Sciences*, V250: 89–104.
Gladwin, H., 1970. *East is a Big Bird.* Cambridge, Massachusetts, Harvard University Press.
Goodman, R., 1992a. *Lakota Star Knowledge: Studies in Lakota Stellar Theology.* Rosebud, South Dakota, Sinte Gleska University Press, Rosebud Sioux Reservation.
———, 1992b. "On the Necessity of Sacrifice in Lakota Stellar Theology as seen in 'The Hand' Constellation and the Story of the Chief who Lost his Arm," in *Earth and Sky: Visions of the Cosmos in Native American Folklore*, ed. R. Williamson and C. Farrer. Albuquerque, University of New Mexico Press, 215–20.
Gove, P., 1993. *Webster's Third New International Dictionary of the English Language, Unabridged.* Springfield, Massachussetts, Merriam-Webster.
Griffin-Pierce, T., 1992. *Earth Is My Mother, Sky Is My Father: Space, Time, and Astronomy in Navajo Sand Painting.* Albuquerque, University of New Mexico Press.
Heidel, A., 1942. *The Babylonian Genesis.* Chicago, University of Chicago Press.
Henry, T., 1928. "Ancient Tahiti," *Bishop Museum Bull.* 48.
Heyen, G., 1962. "Primitive Navigation in the Pacific-1," In *Polynesian Navigation*, ed. J. Golson. Wellington, Polynesian Society, 64–85.
Hilder, B., 1959. "Polynesian Navigational Stones," *Journal of the Institute of Navigation* 12(1): 90–97.
Hines, T., 1991. "Biorhythm Theory: A Cortical Review," in *Paranormal Borderlands of Science*, ed. K. Frazier. Amherst, New York, Prometheus.
Hocart, A., 1941. *Kingship.* London, Watts.
Hoyle, F., 1981. "The Universe: Past and Present Reflections," *University of Cardiff Report* No. 70.
Hugh Jones, S., 1982. "The Pleiades and Scorpius in Barasana Cosmology," in *Ethnoastronomy and Archaeoastronomy in the American Tropics*, ed. A. Aveni and G. Urton, and *Annals of Academy of Sciences* 385: 183–202.
Johnson, D., 1998. *Night Skies of the Aboriginal Australians: A Noctuary.* Sydney, University of Sydney Press.
Jonnes, J., 2003. *Empires of Light: Edison, Tesla, Westinghouse and the Race to Electrify the World.* New York, Random House.
Kerr, J., 1989. *The Mayan Vase Book*, vol. I. New York, Kerr Associates.
Kohn, R., 1981. *The Shape of Sacred Space: Four Biblical Studies.* Chico, California, Scholars Press.
Krupp, E. C., 1978. "Astronomers, Pyramids and Priests," in *In Search of Ancient Astronomies.* New York, Doubleday, 203–39.
———, 1989. "The Cosmic Temples of Old Beijing," in *World Archaeoastronomy*, ed. A. Aveni. Cambridge, Cambridge University Press, 65–74.
———, 1997. *Skywatchers, Shamans, and Kings: Astronomy and the Archaeology of Power.* New York, Wiley.
———, 2000. "Sky Tales and Why We Tell Them," in *Astronomy Across Cultures: The History of Non-Western Astronomy*, ed. H. Selin. Dordrecht, Kluwer, 1–30.
Kursh, C. and T. Kreps, *Starpaths: Linear Constellations in Tropical Navigation*, (unpublished manuscript).
Langdon, C., J. Fotheringham, and C. Scoch, 1928. *Venus Tablets of Ammizaduga.* Oxford, Oxford University Press.
Lee, T. (ed.), 1985. *Los Codices Maya.* San Cristobal, MX, Univ. Autonoma de Chiapas.

Lewis, D., 1974. "Voyaging Stars: Aspects of Polynesian and Micronesian Astronomy," in *The Place of Astronomy in the Ancient World*, ed F. Hodson, Philosophical Transactions of the Royal Society. London, A 276: 133–48.

———, 1976. *We the Navigators.* Honoluolu, University of Hawaii Press.

Lippincott, K. (ed.), 2000. *The Story of Time.* London, Merrell Holberton.

Loewe, M. and C. Blacker (eds.), 1981. *Divination and Oracles.* London, George Allen and Unwin.

Lopez Austin, A., L. Lopez Lujan, and S. Sugiyama, 1991. "The Temple of Quetzalcoatl at Teotihuacan, Its Possible Ideological Significance," *Ancient Mesoamerica* 2(1): 93–105.

Louis, R., 1975. *Child of the Hogan.* Provo, Utah, Brigham Young University Press.

Lovi, G. and W. Tirion, 1989. *Men, Monsters and the Universe.* Richmond, Virginia, Willman-Bell.

MacDonald, J., 1998. *The Arctic Sky: Inuit Astronomy, Star Lore, and Legend.* Toronto, Royal Ontario Museum and Nunavut Research Institute.

Magaña, E., 1988. *Oríon y la Mujer Pléyades: Simbolismo Astronómico de los Indios Kaliña de Surinam.* Dordrecht, Foris.

———, 1996. "Tropical Tribal Astronomy: Ethnohistorical and Ethnographic Notes," in *Songs from the Sky: Indigenous Astronomical and Cosmological Traditions of the World*, ed. V. del Chamberlain, J. Carlson, and M. J. Young. Bognor Regis, Ocarina Books, 244–63.

Makemson, M., 1938. "Hawaiian Astronomical Concepts," *American Antiquity* 40: 370–83.

Mann, B., 2000. *Iroquoian Women, the Gantowisas.* New York, Peter Lang.

Marshack, A. 1972 *The Roots of Civilization.* New York, McGraw-Hill.

———, 1989. "North American Calendar Sticks: The Evidence for a Widely Distributed Tradition," in *World Archaeoastronomy*, ed. A. Aveni. Cambridge, Cambridge University Press, 308–24.

Maurtua, V. [Anonymous Chronicler], 1906. "Discurso de la Sucesión i Gobierno de los Yngas," in *Juicio de Limites Entre el Perú y Bolivia*, Prueba Peruana 8. Madrid, Chunchos, 149–65.

Meyer, J., 2001. *Myths in Stone: Religious Dimensions of Washington D.C.* Berkeley, University of California Press.

Milbrath, S., (in press). "The Maya Lord of the Smoking Mirror," in *Tezcatlipoca: Trickster and Supreme Aztec Deity*, ed. E. Baquedano. Boulder, University Press of Colorado.

Miley Theobold, M., 2005. "The Tyme Appointed," *Colonial Williamsburg Foundation*, Williamsburg, Virginia. Autumn, 20–26.

Miller, M., 1986. *The Murals of Bonampak.* Princeton, Princeton University Press.

Morieson, J., 1996. "The Night Sky of the Boorong: Partial Reconstruction of a Disappeared Culture in North-West Victoria," MA Thesis, University of Melbourne.

Moyer, G., 1982. "The Gregorian Calendar," *Scientific American* 246: 144–50.

Murray, W. B., 1982. "Calendrical Petroglyphs of Northern Mexico," in *Archaeoastronomy in the New World*, ed. A. Aveni. Cambridge, Cambridge University Press, 195–204.

Needham, J., 1954–1988. *Science and Civilisation in China.* Cambridge, Cambridge University Press.

———, 1974. "Astronomy in Ancient and Medieval China," in *The Place of Astronomy in the Ancient World*, ed F. Hodson, Philosophical Transactions of the Royal Society. London, A 276: 67–82.

Neugebauer, O. and R. Parker, 1969. *Egyptian Astronomical Texts III: Decans, Planets, Constellations, and Zodiacs.* Providence, Brown University Press, 214–5.

Nissen, H., P. Damerow, and R. Englund, 1993. *Archaic Bookkeeping: Writing and Techniques of Economic Administration in the Near East.* Chicago, University of Chicago Press.

O'Neil, W., 1986. *Early Astronomy from Babylonia to Copernicus.* Sydney, Sydney University Press.

Pankenier, D., 1983. "Early Chinese Positional Astronomy: The Gaoyu Astronomical Record," *Archaeoastronomy, The Bulletin of the Center for Archaeoastronomy* 3: 10–19.

Pannekoek, A., 1961. *A History of Astronomy.* New York, Dover.

Price, D., 1971. *Science Since Babylon.* New Haven, Yale University Press.

Price, S., 1985. "Delphi and Divination," in *Greek Religion and Society,* ed. P. Easterling and J. Muir. Cambridge, Cambridge University Press, 128–54.

Ray, J., 1981. "Ancient Egypt," in *Divination and Oracles,* ed. M. Loewe and C. Blacker. London, George Allen and Unwin, 174–90.

Reiner, E. and D. Pingree, 1985. *Babylonian Planetary Omens.* Malibu, Undena.

Rieu, E. (tr.), 1946. *Homer, The Odyssey.* New York, Penguin.

Rountree, H., 1989. *The Powhatan Indians of Virginia, their Traditional Culture.* Norman, University of Oklahoma Press.

Rowe, J., 1979. "An Account of the Shrines of Ancient Cuzco," *Ñawpa Pacha* 17. Berkeley, California, Institute of Andean Studies.

Roy, A. E., 1984. "The Origin of the Constellations," *Vistas in Astronomy* 27: 171–97.

Sagan, C., 1980. *Cosmos.* New York, Random House.

Sahagun, B. de, 1953. *Florentine Codex: General History of the Things of New Spain, Book 7, The Sun, Moon and Stars, and the Binding of the Years,* ed. A. Anderson and C. Dibble. Santa Fe, New Mexico, School of American Research, and Ogden, University of Utah Press.

———, 1961. *Florentine Codex: General History of the Things of New Spain, Book 10, The People,* ed. C. Dibble and A. Anderson. Santa Fe, New Mexico, School of American Research, and Ogden, University of Utah Press.

———, 1981. *Florentine Codex: General History of the Things of New Spain, Book 2, The Ceremonies,* ed. A. Anderson and C. Dibble. Santa Fe, New Mexico, School of American Research, and Ogden, University of Utah Press.

———, 1993. *Primeros Memoriales,* ed. F. Anders. Norman, University of Oklahoma Press.

Saliba, G., 2006. "Islamic Science at its Best," *Journal for the History of Astronomy* 37(2): 233–8.

Salomon, F. (ed.), 1991 *The Huarochiri Manuscript: A Testament of Ancient and Colonial Religion.* Austin, University of Texas Press.

Schele, L. and D. Freidel, 1992. *A Forest of Kings: The Untold Story of the Ancient Maya.* New York, Wm. Morrow and Co.

Schele, L. and N. Grube, 1997. *Notebook for the XXIst Maya Hieroglyphic Forum at Texas.* Austin, University of Texas Department of Art and Art History, and the Institute of Latin American Studies.

Schleiser, K., 1987. *The Wolves of Heaven: Cheyenne Shamanism, Ceremonies, and Prehistoric Origins.* Norman, University of Oklahoma Press.

Silberbauer, G., 1981. *Hunter and Habitat in the Central Kalahari Desert.* New York, Cambridge University Press.

Simon, M. and N. Grube, 2008. *Chronicle of*

the Maya Kings and Queens, 2nd edition. London & New York, Thames & Hudson.

Stahlman, W., 1970. "The Star Fixed Ages of Man," *Saturday Review*, 10 Jan, 99–109.

Taube, K., 2001. "The Symbolism of Wind in Mesoamerica and the American Southwest," in *Road to Aztlan*, ed. V. Fields and Z. Taylor. Los Angeles, Los Angeles Museum of Art.

Tedlock, B., 1992. *Time and the Highland Maya* (revised edition). Albuquerque, University of New Mexico Press.

———, 1999. "Maya Astronomy: What We Know and How We Know It," *Archaeoastronomy: Journal of Astronomy in Culture* 14(I): 39–58.

Tedlock, D., 1985. *Popol Vuh: The Definitive Edition of the Mayan Book of the Dawn of Life and the Glories of Gods and Kings*. New York, Simon and Schuster.

Thompson, J. E. S., 1972. *A Commentary on the Dresden Codex, A Maya Hieroglyphic Book*. Philadelphia, American Philosophical Society Memoirs, 93.

Thompson, R., 1900. *Reports of Magicians and Astrologers of Ninereh and Babylon in the British Museum*. London, Luzac.

Thulin C. (ed.), 1913. *Corpus Agrimensorum Romanorum Hyginus Gromaticus. Biblioteca Scriptorum Graecorum et Romanum Teubneriana*. Stuttgart, Teubner.

Tozzer, A., 1941. *Landa's Relación de las Cosas de Yucatan*. Cambridge, Papers of the Peabody Museum of American Archaeology and Ethnology, Harvard University, 23.

Turner, V., 1972. "Religious Specialists," in *International Encyclopedia of the Social Sciences*,13, ed. D. Sills. New York, Crowell, Collier and MacMillan.

Tyson, N., 2007. "The Cosmic Perspective," *Natural History* 116(3): 22–7.

Urton, G., 1981. *At the Crossroads of the Earth and the Sky: An Andean Cosmology*. Austin, University of Texas Press

Waterhouse, E., 1622. *A Declaration of the State of the Colony and Affaires in Virginia*. London, Virginia Company of London.

Wedel, W., 1977. "Native Astronomy and the Plains Caddoans," in *Native American Astronomy*, ed. A. Aveni. Austin, University of Texas Press, 131–46.

Weinberg, S., 1988. *The First Three Minutes*. New York, Basic.

Wilbert, J., 1981. "Warao Cosmology and Yekuana Round House Symbolism," *Journal of Latin American Lore* 7(1): 37–72

Wilson, P., 1988. *The Domestication of the Human Species*. New Haven, Yale University Press.

Winkler, Capt., 1901. "On Sea Charts Formerly used in the Marshall Islands, with Notices on the Navigation of Three Islanders in General," *Report of the Smithsonian Institution for 1899*, 487–509.

York, M., 1986. *The Roman Festival Calendar of Numa Pompilius*. New York, Peter Lang.

Zolbrod, P., 1984. *Diné bahané: The Navajo Creation Story*. Albuquerque, University of New Mexico Press.

Acknowledgments

I am indebted to Simon Martin, Michael Coe, Tony Johnson, and Peter Warner for their comments on draft versions of this text; and I offer special thanks to Diane Janney and Lorraine Aveni, and to Colin Ridler, Sophie Mackinder, Rowena Alsey, Celia Falconer and the rest of the staff at Thames & Hudson for seeing me through day-to-day labours pertaining to this work.

Sources of Illustrations

Half-title Belli, S. (1569) *Libro del misura con la vista.* Venice. **Frontispiece** British Library, London. **1** Courtesy Anthony Aveni. **2** Rollout K1226 © Justin Kerr. **3** Heidel, A. (1942) *The Babylonian Genesis* (fig. 7). Chicago: University of Chicago Press. **4** Courtesy of T. Griffin Pierce and the University of New Mexico Press. **5** Courtesy of Willmann-Bell Inc. **6** Detail from maps 5 and 6, from Norton, A. P. (1950) *Norton's Star Atlas*, (11th edition). London: Gail and Inglis. **7** Courtesy of Sinte Gleska University, Misson, South Dakota. **8** Photo Patrimonio Nacional, Madrid. **9** Drawing Kit Schweitzer. Courtesy of G. Ammarell. **10** Stahlman, W. (1970) *The Star Fixed Ages of Man*, Saturday Review, 10 Jan, p. 90. **11** From J. Needham, J. (1974) 'Astronomy in Ancient and Medieval China' (fig. 6), in *The Place of Astronomy in the Ancient World*, ed. F. Hodson. London: Philosophical Transactions of the Royal Society. **12** Courtesy American Philosophical Society. **13** Bibliothèque Nationale, Paris. **14** From Hugh-Jones, S. (1982) 'The Pleiades and Scorpius in Barasana Cosmology' (fig. 1), *Ethnoastronomy and Archaeoastronomy in the American Tropics*, ed. Aveni, A. and Urton, G. Annals of the New York Academy of Sciences. **15** From Urton, G. (1981) *At the Crossroads of the Earth and Sky: An Andean Cosmology.* Austin, University of Texas Press. Courtesy the University of Texas Press. **16** Courtesy Anthony Aveni. **17** From Hilder, B. (1959) *Polynesian Navigational Stones*, Journal of the Institute of Navigation 12 (1) (fig. 3). London: Royal Institute of Navigation. **18** Drawing Julia Meyerson. Courtesy Anthony Aveni. **19** From Lewis, D. (1974). 'Voyaging Stars: Aspects of Polynesian and Micronesian Astronomy,' (fig. 6) in *The Place of Astronomy in the Ancient World*, ed. Hodson, F. London: Philosophical Transactions of the Royal Society. **20** Reprinted by permission of the publisher from Gladwin, H. (1970) *East is a Big Bird: Navigation and Logic in Puluwat Atoll* (p. 185), Cambridge, Mass.: Harvard University Press. Copyright © 1970 by the President and Fellows of Harvard College. **21** From Winkler, Capt. (1901). *On Sea Charts Formerly used in the Marshall Islands, with Notices on the Navigation of Three Islanders in General* (fig.1), Report of the Smithsonian Institution for 1899. **22** From Gatty, H. (1979) Nature is Your Guide: How to Find Your Way on Land and Sea' New York: Dover Publications. **Box p. 74** Courtesy Anthony Aveni. **Box p. 78** From Aveni, A. (1997) *Stairways to the Stars: Skywatching in Three Great Ancient Cultures* (fig. 2.10), New York: Wiley Press. Courtesy Wiley Press. **23** Department of the Environment. **24** Courtesy Anthony Aveni. **25** This article was published in Advances in Archaeological Method and Theory 4, 16, Aveni, A. (1981) 'Archaeoastronomy' © Elsevier. **26** Aveni, A. and Mizrachi, Y. (1998) *The Geometry and Astronomy of Rujm el-Hiri, a Megalithic site in the Southern Levant*, Journal of Field Archaeology 25, 475-498. **27** From Aveni, A. and Mizrachi, J. (1998) *The Geometry and Astronomy*, Journal Field Archaeology, 25. Reproduced from Journal of Field Archaeology with the permission of the Trustees of Boston University. All rights reserved. **28** Drawing Kit Schweitzer. Courtesy G. Ammarell. **29** Courtesy Anthony Aveni. **30** Courtesy of T. Griffin Pierce and the University of New Mexico Press. **31** Photo The Field Museum, Chicago, neg. no. A106901c. **32** Barbara Brandli. Courtesy Anthony Aveni. **33** Makemson, M. (1941) *The Morning Star Rises*, fig. 4, p. 109. Yale University Press. **34** Courtesy Anthony Aveni. **35** Photo Mary Miller (inset: Courtesy Anthony Aveni). **36** Rob Wood/Wood Ronsaville Harlin, Inc. Courtesy Anthony Aveni. **37** Kenneth Garrett. Courtesy Anthony Aveni (inset: Courtesy of D. Carrasco, Mesoamerican Archive, Harvard University). **38l** Courtesy Antiquity Publications Ltd. **38r** Piacenza Museum, Italy. **39** From Frigerio, P. (1933). *Antichi Istrumenti Technici: a proposito di una stela funeria romana del Museo di Como.* Como: C. Nani. **40** From Needham, J. (1974) 'Astronomy in Ancient and Medieval China' (fig. 5), in *The Place of Astronomy in the Ancient World*, ed. Hodson, F. London: Philosophical Transactions of the Royal Society. **41** The Colonial Williamsburg Foundation. **Box p. 163** Drawing P. Dunham. Courtesy Anthony Aveni. **42** Vorderasiatisches Museum, Staatliche Museen zu Berlin/bpk Berlin, photo Karin März. **43** From Schlesier, K. (1987) *The Wolves of Heaven: Cheyenne Shamanism, Ceremonies, and Prehistoric Origins* (fig. 1). Norman: University of Oklahoma Press. Eugen Diederichs. **44** Courtesy of Sinte Gleska University, Misson, South Dakota. **Box p. 195** National Museum of American History, Smithsonian Institution, Washington D.C. photo no. 89-4591. **Box p. 206** Courtesy Simon Martin. **Box p. 208** Courtesy Simon Martin. **45** Musée du Louvre, Paris. **Box p. 221** Courtesy Anthony Aveni. **46** Hans Blohm/Masterfile. Courtesy Anthony Aveni. **47** Courtesy Wiley Press. **48** Courtesy Wiley Press. **49** Courtesy Wiley Press. **Plates: 1** Alexander Marshack. Courtesy Anthony Aveni. **2** Courtesy Anthony Aveni. **3** From Blier, S. (1987) *The Anatomy of Architecture.* New York: Cambridge University Press. Courtesy S. Blier. **4** Photo 2007 Peabody Museum, Harvard University 48-63-20/17561. **5** Rollout K1196 © Justin Kerr. **6** Detail from the Dresden Codex. Sächsische Landesbibliothek, Dresden. **7** From Diaz, G. and Rogers, A. (1993) *The Codex Borgia* (pl.10), Dover Publications. Apostolic Library of the Vatican. **8** From Diaz, G. and Rogers, A. (1993) *The Codex Borgia* (pl. 17), Dover Publications. Apostolic Library of the Vatican. **9** Museo delle Tavolette di Biccherna, Archivio di Stato di Siena.

Index

Page numbers in *italic* refer to illustrations, numbers in **bold** refer to color plates.